UNDERSTANDING BIODIVERSITY

UNDERSTANDING BIODIVERSITY

DAVID ZEIGLER

Westport, Connecticut
London

Library of Congress Cataloging-in-Publication Data

Zeigler, David, 1950–
Understanding biodiversity / David Zeigler.
    p. cm.
    Includes bibliographical references and index.
    ISBN 978–0–275–99459–4 (alk. paper)
    1. Biodiversity.   I. Title.
QH541.15.B56Z45   2007
577—dc22   2007000072

British Library Cataloguing in Publication Data is available.

Library of Congress Catalog Card Number: 2007000072
ISBN-13: 978-0–275–99459–4
ISBN-10: 0–275–99459–7

First published in 2007

Praeger Publishers, 88 Post Road West, Westport, CT 06881
An imprint of Greenwood Publishing Group, Inc.
www.praeger.com

Printed in the United States of America

The paper used in this book complies with the
Permanent Paper Standard issued by the National
Information Standards Organization (Z39.48–1984).

10 9 8 7 6 5 4 3 2 1

Copyright Acknowledgments

The author and publisher gratefully acknowledge permission for use of the
following material:

Extracts from the Louis MacNeice poem "Snow" are reprinted with permission of
David Higham Associates Limited.

There are more things in heaven and earth, Horatio,
Than are dreamt of in your philosophy.
                              Shakespeare, *Hamlet*

World is crazier and more of it than we think,
Incorrigibly plural. I peel and portion
A tangerine and spit the pips and feel
The drunkenness of things being various.
                    Louis MacNeice, Snow—*Collected
                    Poems*, Faber and Faber, 1979

Contents

Preface

It (The Tree of Life) is surely a being that is greater than anything any of us will ever conceive of in detail worthy of its detail.
— Daniel Dennett, *Darwin's Dangerous Idea*, 1995

The earth has spawned such a diversity of remarkable creatures that I wonder why we do not all live in a state of perpetual awe and astonishment.
— Howard Ensign Evans, *Life on a Little-Known Planet*, 1968

Let me begin by saying that the title of this book is also its goal or purpose. My hopes are that the reader will gain a better understanding of the diversity of life on this special planet we call home. At the same time, another goal is to illustrate that in reality one cannot fully understand or comprehend the biodiversity of our planet (as the quotes given earlier imply). I hope the reader will be moved in both directions simultaneously, gaining a fuller appreciation and understanding of earth's biodiversity, while at the same time building the argument that the earth's biodiversity is so vast, complex, and multileveled that no one could possibly comprehend it all. There is nothing unusual or necessarily contradictory about these two goals. Science has as its goal the complete objective understanding of all physical phenomena in the universe. All scientists know this to be an unachievable goal, but one worth pursuing nonetheless—progress without an attainable end. This is something like the familiar philosophical idea that life is a journey, not a destination. In the present case, progress toward understanding biodiversity is an enriching pursuit, even if full comprehension is beyond our reach.

The term "biodiversity" is a relatively new hybrid of the two words "biological diversity." This shorthand term has taken on a life of its own in the last few years and has become a buzz-word for those interested in, and concerned about, the variety of life on earth. But just what is biodiversity? There are many definitions out there, even from the biological community, some seemingly more restrictive than others. The more common definition will be further elaborated in Chapter 1, the one that emphasizes three aspects of biodiversity: species diversity, genetic diversity, and ecosystem diversity. Some definitions include one or two other parameters of biodiversity such as "higher-level" diversity between taxonomic orders, classes, phyla, etc. (the topic of Chapter 5). The main thrust of this book is that biodiversity includes much more, as revealed in this line by Terry Erwin: "Biological diversity (biodiversity) is in fact the product of organic evolution, that is, the diversity of life in all its manifestations." Biodiversity includes all detectable parameters along which we find diversity in living organisms, and most of these have been little mentioned in the majority of writings on biodiversity. This book will touch on several (though certainly not all) of these overlooked areas of diversity in an attempt to move the reader toward a better understanding and appreciation of the full sweep of earth's biodiversity. Though frequently used, the term biodiversity is a relatively new term that just happens to refer to the most varied and complex set of phenomena known to exist in the universe. That seems reason enough to devote a book to more fully explaining its meaning.

Most of our detailed knowledge of living things is based on studies of a very small minority of the life forms known to exist. Most of the earth's species have not even been discovered and recognized by scientists. Of those species that have been discovered and named (approaching two million), all we know about the great majority is that they do exist. There are almost endless possibilities for continued learning if we stopped discovering new species (and who would want to?) and started trying to fully understand those now known to exist.

Anyone who desires to be a knowledgeable well-rounded biologist has entered into a quest, which is near impossible (though thoroughly rewarding and worthwhile!) due to the vastness of the subject matter. It's the old story of "the more you know, the more you know you don't know," as related in this famous quote: "We live on an island of knowledge surrounded by a sea of ignorance. As our island grows, so does the shore of our ignorance" (John A. Wheeler, Quoted in *The New Challenges. Scientific American*, Dec. 1992).

Knowledge in the biological sciences has grown so immensely over the past few decades that it is increasingly difficult to keep up with even the major discoveries. Many of these discoveries have increased our understanding of the range and depth of the earth's biodiversity. Only rarely does a discovery reveal less diversity than had been suspected. One such example comes from the recently completed human genome project. Previous estimates of the number of human genes ranged from around 70,000 up to 150,000, but the results of the human genome project now indicate around 25,000 genes in the human genome. Genetic diversity is one of the major levels or components of total biodiversity, and at least in this one instance, new findings set a lower number for human genetic diversity than had been suspected. Such an example is the rare exception to the general trend of discovering more biodiversity than had ever been dreamed of in the variety of life's parameters.

Though full comprehension of biodiversity is probably impossible, surely everyone should invest some effort toward understanding the awesome diversity of life on earth. Only then can we know how special this world actually is, how valuable it is, and how worthy of our awe and protection it is. Though there is much speculation about life on other worlds by scientists, writers, and the man on the street, it is all so far just that—speculation. Right here "under our noses" lies the most amazingly complex, beautiful, bizarre, and awesome collection of life known to exist anywhere in the universe, and most of it is still undiscovered.

Many believe that you are what you know. Like E. O. Wilson, most biologists accept a basic underlying human nature resulting from genetic influences, but clearly most individual human knowledge is culturally or environmentally acquired. What you are able to learn and understand about the world you inhabit really does shape who you are, and one who knows and appreciates nothing of the wonders of this living world is a very poor person indeed—no matter the state of their bank account. One of Darwin's most often quoted lines comes at the end of *The Origin of Species* where he writes: "There is grandeur in this view of life, with its several powers, having been originally breathed into a few forms or into one; and that whilst this planet has gone on according to the fixed law of gravity, from so simple a beginning endless forms most beautiful and most wonderful have been, and are being evolved."

Darwin seems to be saying that the grandeur is not only in the continuity of life through evolution, but also in the "endless forms most beautiful and most wonderful." Darwin was a very fortunate person in that during his 5-year voyage around the world he was able to see first-hand more of the earth's biodiversity than the vast majority of biologists (or people in general) who have ever lived. As a result of his experiences and his understanding of natural selection, Darwin was probably the first person to feel and understand what he as a human actually was, and what a wonderfully diverse and strange world we inhabit. He also knew that he and humanity were truly a connected part of this living world—a single thread in the fabric of life. This knowledge does and must deeply affect anyone who understands it. Such understanding transforms one's self image, ego, sense of priorities, and sense of beauty and meaning such that one simply cannot be the same person they were before these facets of reality were incorporated into their consciousness.

Though this small book cannot possibly suffice to reveal the immensity of biodiversity and what it may mean to us, my hope is that it will serve as a partial guide and map which will make the reader more likely to continue on a quest of discovery and understanding of this life-filled planet we call home. Some people, both scientists and nonscientists alike, believe that science should ultimately be concerned with benefiting human society in some tangible way, such as the role of science in the health sciences. Some scientific knowledge can also be applied in the area of technology toward solving many diverse human problems (or perceived problems). But many scientists would argue (as would others) that aside from solving problems, scientific knowledge has value in and of itself because of the effect it can have on our personal understanding and appreciation of the world and universe in which we live.

If I did not know that microorganisms existed widely and in tremendous variety, or that the sun was but one star among trillions (in billions of galaxies), or that I am directly related to every living organism on this planet, or hundreds of other amazing truths concerning this world, then I would be a very different person or "self." I would, I feel, be intellectually crippled by not knowing these things. The world could not possibly be as interesting, awesome, and personally "real" without this knowledge.

Scientific knowledge therefore contributes to personal enlightenment, and this benefit should certainly be recognized as one of the most important products or outcomes of scientific endeavor. Unfortunately, this enriching knowledge is still possessed by far too few of our fellow humans, many of whom would approach their relationship to nature, and to life in general, quite differently (almost certainly more wisely—especially in terms of their concern for biodiversity) if they had more of this knowledge. Those without even a minimal store of this type of knowledge are experiencing a very different, more impoverished, and far smaller universe than the one which science has illuminated.

1

The Living Planet: The Unity and Diversity of Life

What characterizes the living world is the basic unity that underlies its tremendous diversity.

—Francois Jacob, *The Possible and the Actual*, 1982

A blue algae, an infusorian, an octopus, and a human being—what have they in common? With the discovery of the cell and the advent of cellular theory, a new unity could be seen under this diversity.

—Jacques Monod, *Chance and Necessity*, 1971

The term "life" encompasses the totality of life on this planet. The term seems obviously to refer to some commonality which all living things share. We also use the word "biodiversity" to refer to the life of our planet, and this term obviously refers to and stresses the diversity of earth's life-forms. These two terms each imply something different about the life of this planet. Indeed, life does possess these two seemingly disjunct qualities of unity and diversity. This book will largely be addressing the diversity of life, but this first chapter will briefly focus on the unity of life by looking at some of the universal features of life on earth.

THE UNITY OF LIFE

The common ancestry of living things has resulted in a commonality among the Earth's organisms (Daniel Chiras, *Biology: The Web of Life*, 1993).

Life can be said to show unity for a number of reasons. The first and primary reason stems from the very definition of life. That is, life is defined by a set of *common* characteristics or properties which are shared by all life-forms. Though some biologists seem to have a hard time coming to grips with a pragmatic definition of life, there are varying "standard" lists of characteristics which many feel can be used to define life. These lists can be found in almost any introductory biology textbook. Certainly, exceptions can be found to some of these characteristics, especially if one argues that viruses are organisms (most biologists don't accept that they are). Some of these characteristics may be more essential than others, but they are all valid, and *collectively* they serve as a very accurate description of what it means to be a living organism. Listed below are 12 of the characteristics which are shared by life-forms as diverse as bacteria, pine trees, bread mold, and zebras.

1. *Reproduction:* All species of living things must be able to reproduce or they would eventually go extinct. This is because all individual organisms eventually die. Some bacteria may live as individuals for only 20 minutes or so (though they may not actually die after this time, only divide to form two "new" individuals). Most animals (the majority being insects) live only a year or less. A few trees, such as the bristlecone pine can live for more than 4,000 years. Even in bacteria the range of life spans is believed to be enormous, from those that live out their individual lives in 20 minutes to those which are believed to live slow lives of 100 years or more in the cold depths of the earth or deep in arctic ice sheets. But even the 4,000-plus-year life of a bristlecone pine is but a moment in evolutionary and geologic time. Many species survive from one to several million years, so surely reproduction is occurring in those species. Actually, not all *individuals* are capable of reproducing offspring. Worker ants and honeybees are sterile and have no ability to mate or reproduce offspring. What they are capable of is increasing their "genetic fitness" by perpetuating copies of their genes, and we take up this more complex matter next.

2. *Biological selfishness:* This characteristic is one of the most essential in any definition of life, though it is still not included in many general biology textbooks. Several workers contributed to this important concept, but it was Richard Dawkins who elaborated the point most clearly in his revolutionary book *The Selfish Gene.* The key idea here was that we finally understood what life was all about, that is, we came to clearly see that all living organisms have one overriding common goal. That goal is to perpetuate the flow of *copies* of their own genes into the next generation, typically by producing offspring. This goal and all efforts by organisms to accomplish it are referred to as biological selfishness. Even the sterile worker bees and ants are involved in this goal because by working for the good of their queen and the hive, they insure that their queen mother will continue to lay eggs which will be their sisters. These new sister workers will be related to current workers by 25 percent (if they had

different fathers) or by 75 percent (if they had the same father). In short, the workers are striving to increase the survival of copies of their genes in future generations of workers, but not by producing their own offspring.

The reason this goal is referred to as selfish is to contrast it with an older idea which was generally accepted by nearly everyone. This older idea was that organisms reproduced (and did a lot of other things) for the "good of their species." It was said that they reproduced to perpetuate their species. This idea has been effectively dead and obsolete since the publication of *The Selfish Gene* in 1975. No individual organisms (except for some humans) do anything for the continuation or good of their species; rather, everything they do is directed toward promoting the survival of copies of their own individual genes in future generations.

Promoting your own genes is most definitely not the same thing as promoting your species. In many cases, being biologically selfish will lead to the outright killing of fellow species members. Male lions, having just won control of a pride of females (from previous controlling males), will often succeed in killing any small cubs which have not been weaned, and which were fathered by the previous males. This will allow those nursing females to quickly come into heat so that the new males can father their own offspring. This is admittedly biological selfishness of a relatively extreme form, but it is just as obviously not a behavior that could be construed as concern for the species as a whole.

3. *All life is cellular:* Living organisms are composed of structural units called cells. This universal property is referred to as "The Cell Theory." Though there are many variations on cell structure across the five kingdoms of life, all life is cellular in structure. There are indeed some tissues in some organisms which are not clearly cellular (because there are no cell divisions between numerous nuclei) such as the epithelial tissue of rotifers or the "plasmodium" in slime molds, but at least somewhere in the complete life cycle of any living species, there is a cellular stage, even if it is only a sexual gamete or a zygote (fertilized egg).

4. *Movement:* This is an easy one. Since all life is cellular, there is *purposeful* directed movement, at least within the cells. Examples would be the movement of nutrients into cells, the movement of wastes out of cells, movement of messenger RNA out of the nucleus, movement of new proteins from the endoplasmic reticulum to the golgi bodies, etc. Cells cannot survive without these movements, so movement is a characteristic of life. In multicellular organisms there is often movement of water and sugars through vascular tissue, movement of blood through a circulatory system, movement

of cytoplasm in the hyphae of some molds, etc. Of course many organisms can also move through their environment by a variety of methods such as swimming, crawling, walking, flying, etc.

5. *Growth:* All living organisms undergo a process of growth. Bacteria and many of the protists can divide asexually by a process called fission. Fission typically results in two individuals that are only half the volume of the "parent" individual. The two new individuals then feed and grow until they are mature and ready to undergo fission themselves. This growth between divisions is referred to as *cellular growth*, since the organism itself is a single cell. Multicellular organisms (plants, fungi, animals, and some multicellular algae) also have cell divisions, but the new cells remain a part of the same "body." The result is that the organism grows *by the addition of more cells* to its body structure. Adult humans contain many more cells than are present in a newborn baby of only seven pounds. Multicellular organisms grow by both the addition of more cells and by cellular growth.

6. *Metabolism:* Metabolism consists of all the directed and purposeful chemical reactions that occur in living cells, many of which are controlled by enzymes. Some of the reactions are anabolic in which large molecules are built up from smaller molecules as occurs in growth, development, and the production of cell products. Other reactions are catabolic in which large molecules are broken down into their subparts as in digestion, absorption of tissues, or the respiration of sugars. All living cells undergo some characteristic combination of chemical reactions which collectively make up the metabolism of that cell. Of course different cells, even in the same multicellular organism, have different total metabolisms. A fat cell's metabolic activities would vary widely from those of a neuron (though there would be some overlap). Metabolisms also vary widely between different species, especially in distantly related groups such as plants and animals (more on metabolic diversity in Chapter 10). Even so, there are great similarities in the metabolisms of even distantly related organisms.

7. *Irritability:* Irritability is defined as the ability to respond to stimuli. All living organisms have the ability to respond in some measurable way to at least some stimuli. Stimuli are conditions or changes in the conditions of the immediate environment or interior of the organism that evoke a response. Plant roots grow down in response to gravity, while their stems grow up. A human muscle cell will respond to insulin by removing glucose from the blood and storing it as glycogen (unless the person is diabetic). *Euglena*, a photosynthetic unicellular organism, will use its flagella to swim

toward brightly lit areas where it can photosynthesize, and it will avoid shaded areas. From the level of cells to the level of organisms, all life shows irritability to at least some stimuli.

8. *Life evolves:* Life got started in some way (which we'll probably never know precisely) around 4,000,000,000 years ago on our planet, and has been evolving and diversifying ever since. Even modern bacteria have evolved in countless ways (mostly biochemical) since the earliest bacteria. Bacteria continue to be among the most rapidly evolving organisms on the planet, as demonstrated by the evolution of antibiotic resistance by several disease-causing bacteria over the last 50 years. There are a few species which seem to have almost ceased evolutionary change such as the horseshoe crab and the cockroach. Though these and others have remained relatively unchanged for many millions of years according to the fossil record, they have undoubtedly changed in some internal or biochemical features which we have no way to detect. In contrast, other groups have undergone relatively rapid change and diversification. Our own family, the Hominidae, as revealed by the fossil record, has shown rapid change and diversification over the past 5 million years. Another example of extremely rapid evolution and speciation would be the fruit flies of the Hawaiian Islands. The Hawaiian island chain was formed (and continues to form) through volcanic activity, and is less than 10,000,000 years old. In this short time (short in geologic and evolutionary terms) one or two colonizing species of fruit flies have evolved and speciated into over 700 distinct species, some with very diverse morphologies and behaviors.

9. *Life uses DNA as its genetic "information" molecule:* It is truly amazing that both bacteria and humans (and all the other organisms) use DNA (deoxyribonucleic acid) to encode the blueprints for their bodies, their physiologies, and their behaviors. There are some viruses whose genetic material is in the form of RNA, but, as mentioned earlier, viruses are typically not given the status of living organisms because they lack many of the characteristics of life in this list (such as growth, independent metabolism, cellular structure, etc.). DNA is made up of only four different building blocks known as nucleotides, yet this four-letter chemical alphabet can spell out (in very long sequences) all the information needed to construct and control the millions of species of living organisms and most of their interactions with their environments. Though "nature" does not totally control animal behavior, even the "nurture" aspect of some animal lives would have no effect without a material body and nervous system, neither of which would exist functionally without the blueprints contained in the DNA.

10. *Life uses ATP as its energy "transfer" molecule:* Since all life carries out some type of metabolism (#6), and since part of that metabolism requires the directed capture and use of energy, there must be some physical way to do so. Though there are a few other molecules which can capture and release usable cellular energy, ATP (short for adenosine triphosphate) is by far the most common molecule serving this purpose, and all organisms capture at least some of their usable energy in ATPs. When any organism metabolizes energy-rich molecules, one goal is to capture some of the energy in a usable form that can run various cellular jobs. Sugars and other nutrient molecules contain a great deal of energy in their chemical bonds, but cells cannot run directly on these energy-rich molecules because they contain far too much energy for any one cellular job. They must be "changed" into smaller amounts of energy currency (like changing a 20-dollar bill into quarters for use in vending machines which take only quarters). In this analogy, the quarters are the ATP molecules. They contain just the right amount of energy for most individual energy-requiring jobs in the cell. In short, all known life-forms create and use ATP molecules in their metabolic processes.

11. *Life is composed of a unique combination and ratio of elements:* This may seem like a rather picky characteristic from a biological perspective, but a chemist or biochemist would not think so. Over 98 percent of the mass of any life-form would consist of hydrogen, carbon, oxygen, and nitrogen, with a pinch of phosphorus and sulfur thrown in. This is a distinctly different list from that of the earth's crust which is 90 percent oxygen, silicon, and aluminum. It is also distinctly different from the earth's atmosphere which is 99 percent nitrogen and oxygen. The particular list and ratios of life's elements is in a real sense a "chemical signature" of life as we know it.

12. *Life is characterized by low levels of entropy:* In physics, entropy is a term referring to the level of "disorder" in a system. If a system has a high level of entropy, it is very disordered, simple, random, and nonstructured. If a system has a low level of entropy, it is ordered, complex, nonrandom, and highly structured. Even the simplest bacterial cell is a highly ordered and nonrandom arrangement of molecules that work together in very specific and complex ways. Eukaryotic cells are even more structured, having numerous organelles and membranous structures. Complex multicellular organisms raise structure and complexity by at least an order of magnitude over the unicellular organisms. We will discuss these and other levels of living structure in Chapter 2. A bacterial cell, which is too small to be seen, has much more structure and complexity than the moon! The moon, though very big, is just a large collection of relatively

simple molecules held together in a nonstructured mass by gravity. In contrast, a bacterial cell, or even a bacterial DNA molecule, is a very highly structured arrangement of molecules bonded in very nonrandom ways. This is much like saying that a jumbled pile of bricks, no matter how massive, does not have the structure of a single brick "structure" such as a chimney or planter box.

This review of life's characteristics reminds us of the broad underlying unity of life on earth. Though this unity is impressive and real, there can be great diversity within many of those characteristics, both within and across species. Chapter 10 will illustrate some of the great diversity in metabolism (#6) that exists among groups of organisms. Chapter 12 will do the same for reproduction (#1), and Chapter 14 will cover behavioral diversity (a subcategory of irritability, #7). Though all life is cellular (#3), a vast diversity of cells has evolved on this planet, as suggested by this very cursory listing:

- Prokaryotic cells (bacteria)—Eukaryotic cells (plant and animal cells).
- Motile cells (*Paramecium*, sperm)—nonmotile cells (xylem, neurons).
- Cells with contractile vacuoles—cells lacking contractile vacuoles.
- Cells capable of independent life (*Amoeba*)—cells incapable of independent life (neurons).
- Contractile cells (muscles)—noncontractile cells (xylem).
- Nucleated cells with mitochondria (liver cells)—nucleated cells lacking mitochondria (*Giardia*—a protist parasite).
- Short-lived cells (sperm & red blood cells)—long-lived cells (neurons).
- Cells with cell walls (plant & fungal cells)—cells lacking cell walls (animal cells, *Amoeba*).
- Haploid cells (gametes, all cells in a male wasp)—Diploid cells (zygotes, body cells of most animals and "higher" plants).
- Cells capable of endocytosis (macrophages, *Amoeba*)—cells incapable of endocytosis (muscle cells).

Again and again, and at every level, we can see both unity and diversity in the life of our planet. Another major factor contributing to the unity of life is common descent. Darwin came to understand (as do scientists today) that all life-forms could trace their ancestry back to one original life-form. In short, unity is in large part inherited from common beginnings. In outward appearance, the earth's life-forms seem to display a tremendous amount of diversity, but under those surface appearances lays a substantial unity resulting from common descent.

Most of the 12 characteristics reviewed above are certainly due to common descent, as are a great many more. Common descent leads to a certain universal unity of all life, but it leads to even greater levels of unity within closely related groups. Though 20 percent of all known animal species are beetles, which illustrates great diversity in this animal order, they are nevertheless all beetles! That is, these thousands of beetle species share a great many characteristics in common; characteristics that group them (unify them) as beetles and make them distinct from ants, flowering plants, or ciliates. All flowering plants share a different set of unifying characteristics, as do all Kingdom groupings, phyla or division groupings, class groupings, etc.

This underlying unity resulting from common descent was obvious to Darwin, and it is still obvious to anyone who surveys in depth the life of this planet. However, the outward surface diversity is often more obvious and striking to the eye than the underlying unity. When we look at organisms, we see only the surface of that life-form. We do not see the numerous similarities at the several levels underlying that surface. Actin (probably the most common protein in animals) is basically the same molecule in all animals and protists. Cilia and flagella are basically the same organelle whether they are found in animal cells or on the cell of a protist such as *Paramecium.* Striated skeletal muscle is basically the same tissue in structure and function in the great majority of all animal species. Actin, cilia, and striated muscle are, respectively, known as homologies within the groups that share them.

Homologies, or homologous "structures," are those structures that are shared among several related species of organisms due to common descent. Though homologies can end up looking quite different in different species, such as the canine teeth of a human versus the canine teeth of a walrus (the tusks), most homologies still look very similar, if not indistinguishable, like the flagella or striated muscle examples above. One would have to do an analysis at the molecular level in order to distinguish chimpanzee skeletal muscle from human skeletal muscle since the gross morphology of the two is identical. Take the hair and skin off a chimpanzee and a human, and you would have a remarkable similarity of underlying body appearance. There are at least thousands of examples of homologies within most related groups. They exist from the level of molecules and cells up to the level of tissues, organs, and gross morphology. All these homologies contribute to a certain level of unity within those related groups. It seems to be a part of our human nature that we emphasize and dwell on differences while downplaying similarities. Racism is one of the unfortunate results of this common human tendency. If we have some difficulty in seeing the unity of our own species, how much less are we able to see and appreciate the unity of life's larger groupings?

In order to *truly* understand and appreciate life, not just scientifically—but from any valid perspective, it is essential that one delve into and understand this unity arising from common descent. An old criticism from the lay-public of some scientific studies concerns the use of rats in medical research. The criticism is simply that humans are not rats, therefore how could valid medical knowledge come from using rats in cancer research? Though it is true that humans are not rats, and some differences in our cellular and molecular makeup might reduce the validity of some such studies, the fact remains that both rats and humans are mammals, with very similar chemistry, physiology, and genetics, all due to our common descent from a common mammal ancestor. If a substance causes cancer in rats, it may very well cause cancer in humans because rats and humans have similar tissues that might well respond in similar ways to the same insults. Biologists understand perfectly well the underlying unity of mammals that makes such research useful and predictive.

A third factor that contributes to the unity of life is evolutionary convergence. Though there are millions of different species of organisms on this planet, each with at least a slightly different ecological "niche," there are not really millions of distinctly different ways of making a living in the world. Common environments and limitations of the trophic pyramid often foster similar adaptations and ways of life in two or more species that do not share those adaptations as homologies. Plants are related through common ancestry and share photosynthesis as a homology (making them "producers"). Only a few plants have adapted to living in the desert areas of the world. Some of those that have done so have descended from very different ancestral lines and morphologies, but in adapting to the harsh stresses of the desert environment, some have converged on strikingly similar morphological solutions to certain desert stresses. At least a few have converged on what appears to be a cactus-like morphology, including thick stems for water storage, thick epidermal layers to reduce water loss, and spines to protect the water-bearing tissues from grazing animals (Figure 1.1).

Convergence also occurs in nonmorphological characteristics. When biologists speak of "sit-and-wait" predators, they are aware that many animals from many phyla and classes make their living utilizing this

Figure 1.1 One of the classic cases of convergent evolution in desert plants. The plant on the left is a true cactus in the family Cactaceae, a family of around 2,000 species, most of which are adapted to dry conditions. The plant on the right is a spurge, a member of the larger family Euphorbiaceae (7,500 species), most of which look nothing like a cactus and are not desert-adapted. This specimen in the genus *Euphorbia* is one of the few euphorbs which is desert-adapted. It has evolved the thick succulent stems, reduced leaves, and spines typical of most cactus. (*Photo by Leon Jernigan*)

method of hunting (a behavior). Trapdoor spiders, scorpionfish, preying mantids, and leopards can all be accurately described as sit-and-wait predators, all having converged adaptatively on this hunting style. The complex behavioral lifestyle known as eusocial behavior is believed to have evolved separately at least a dozen times in insects, mostly in the bees, ants, and wasps, but also in the relatively unrelated termites.

Another very common lifestyle that numerous species from every kingdom have converged on is that of parasitism, the taking of nourishment from the living bodies of other organisms (their hosts). Not only is this a common method of obtaining nourishment, but it also typically involves many other common adaptations. Perhaps the most common is an increase in the number of offspring (to increase the chances that at

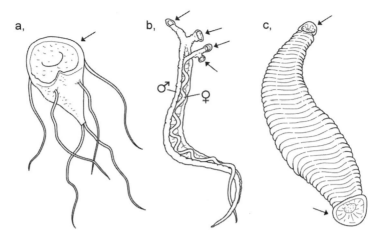

Figure 1.2 The convergent evolution of suckers in three unrelated parasites: (a) *Giardia*, (b) blood flukes, (c) a leech. (*Figure by Jeff Dixon*)

least a few will be able to locate new hosts). Another common parasitic adaptation is the evolution of anchors or holdfast structures used to connect the parasite to the host such that the parasite is not easily dislodged from its hold on host surfaces. Hooks, claws, suckers, and powerful mouth muscles have evolved in parasites with very distant evolutionary relationships. Suckers for example, are found on leeches, flukes, and even the intestinal protist *Giardia* (Figure 1.2).

Though the details of most recognized convergences are not identical, there is enough similarity of form, behavior, and/or lifestyle to justify using the term convergence (though a few biologists are somewhat critical of the concept). Certainly the filter-feeding of a clam is quite different from the filter-feeding of a basking shark. On the other hand, the morphology and gliding behavior of a sugar glider (a marsupial) is remarkably similar to that of a flying squirrel (a placental mammal or eutherian), and this similarity is due entirely to convergence (Figure 1.3). Cases of extreme convergence in form *and* function are admittedly rare. But there are many more general convergences to common "lifestyles" such as predation, parasitism, sociality, polygamy, nectar-feeding (honeybees, hummingbirds, some bats), blood-feeding (bed bugs, leeches, vampire bats, one Galapagos finch), flying (insects, birds, bats, pterosaurs), and others.

There may be other suggestions as to why and how life displays the property of unity, but the three reasons we have covered (definition

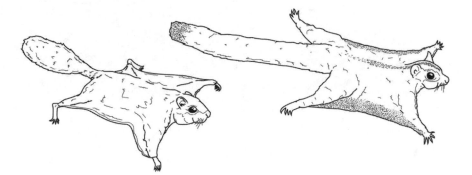

Figure 1.3 A striking case of convergent evolution in form and function between a flying squirrel (left—a placental mammal) and a sugar glider (right—a marsupial). (*Figure by Jeff Dixon*)

of life, common descent, and convergence) are arguably the strongest and most encompassing. One final thought concerning unity is the question—does the unity of life constrain the diversity of life? Since the two properties might be seen as opposites, this seems a logical question, and surely the logical answer to this question is yes. No matter how vast and incomprehensible the biodiversity of our planet may be, it is certainly held within some limits by the shared properties of life, by common descent, and by the convergent effects of limited environmental and lifestyle diversity available on the planet. Those constrains though, at least seem to have been minimal when we survey the vast diversity of earthly life. We will be looking at this diversity in the following pages—diversity so rich and varied that it appears to have overwhelmed the underlying unity, at least at macrolevels of observation.

The Diversity of Life

Biological diversity is the complete array of organisms, biologically mediated processes, and organically derived structures out there on the globe. And it very definitely, as far as I'm concerned, goes way beyond individual species. It goes beyond genetic variance. (Jerry Franklin, Quoted in, *The Idea of Biodiversity*, David Takacs, 1996)

So just what do we mean when we use the term biodiversity? The commonly understood standard meaning among biologists always includes the following three components or levels of diversity:

1. *Species diversity*: This typically refers to a simple count of species that exist on the earth, in a particular ecosystem, in a particular higher

taxonomic grouping such as a phylum, etc. This is certainly the most central and important measure of biodiversity, and it is what most readily comes into the minds of most biologists when they hear the term. This is the topic of Chapter 3.

2. *Genetic diversity*: This refers to the number of different genes and alleles present within a population, a species, even an ecosystem. Genetic diversity can be discussed meaningfully at several levels, and this is the topic of Chapter 6.

3. *Ecosystem diversity*: This refers to the number of distinctly different ecosystems or communities within an area, on a continent, on the earth, etc. In other words a coral reef ecosystem is quite different from the deep abyssal plain, which is quite different from a hydrothermal vent community. This is the topic of Chapter 7.

While a few writers and workers have added a fourth or fifth level to this core meaning, total biodiversity consists of "even more." As the earlier quote by Jerry Franklin suggested, another clear component of biodiversity is "the diversity of ways in which life is diverse." Yes, that statement does make sense—read it again. This idea is occasionally hinted at, but is not typically developed or explained in most sources. The objective of this book is to make a case for, and elaborate on, this idea of multiple levels, scales, and parameters within and along which diversity can and does exist. If biodiversity is wonderfully amazing, how much more so would it seem if we could be more aware of, and understanding of, the multileveled nature of biodiversity.

One does not need to understand the physics of light, color, refraction, etc. to appreciate and comprehend the beauty of a rainbow or a sunset. Some might equally argue that we don't need to better understand or comprehend biodiversity to stand in awe of it, to value it, and to take steps to preserve it. Most scientists would argue in rebuttal that only by knowing more of the specifics of biodiversity could one begin to appreciate the extent, the strangeness, and the beauty of our living world. Only by becoming familiar with some of its details can we ever approach an understanding of how complex and varied the living world actually is. Such knowledge simply affords us more to hold in wonder (which should be desirable). Students often become truly amazed and awed by the workings of the human body *only after* having learned of some of its complexity in the classroom. Would not similar results for biodiversity be likely to occur after learning some of the details of earth's biodiversity?

2

Diversity, Complexity, and Natural Selection

Tapeworms were not put here to serve a purpose, nor were planets, nor plants, nor people. They came into existence not by design but by the action of impersonal natural laws.
— Douglas J. Futuyama, *Science on Trial*, 1995

Evolution carries the fates of organisms where it may, according to the prevailing circumstances, without recourse to any preexisting plan.
— Henry Gee, *In Search of Deep Time*, 1999

The advent of the theory of evolution dispelled any idea of a preestablished harmony imposing a system of relations on organized beings.
— Francois Jacob, *Of Flies, Mice, and Men*, 1998

Actually, "selected" individuals are simply those who remain alive after the less well adapted or less fortunate individuals have been removed from the population. Those who use the word selection should never forget that it really means nonrandom elimination, and that there is no selective force in nature.
— paraphrased from Ernst Mayr, *This is Biology*, 1997

One very natural question that occasionally arises is: Why is life so diverse on our planet? I remember this question from my graduate school days, where it sparked many discussions and several insightful (and some not so insightful) discussions. If the first true organisms were simple bacteria-like cells, and if the goal of organisms is to pass copies of their DNA or genes into the next generation, then why, once that was possible in these early cells, was there further diversification of species and a buildup of greater complexity in some lines?

This question has spawned answers which can and have filled entire books, so this chapter will only attempt to relate a few of the major ideas relating to this important question. To start with an odd but pertinent analogy, if a small town had 50 stores, could all of those stores be donut shops? The answer of course is most surely not. Not everyone in the town would care that much for donuts. More importantly, the population requires other commodities such as eggs, peanut butter, wine, tires, nails, CD players, carpet, detergent, shirts, shoes, batteries, pillows, and a vast number of other things that are not carried by donut shops. The shops of any town must be diversified to serve the community and more importantly to the shop owners, *to survive* financially. Fifty donut shops would create too much *competition* in a small town where only one or two might thrive. This is certainly not a satisfactory analogy to life-forms, but it does illustrate one reason why diversity might develop and be maintained—to avoid excessive competition.

No matter where or how life first evolved on this planet (and we will probably never know this with great certainty), there had to be some variety in the environments available to those early life-forms. If some neighboring environmental space was almost, but not completely, ideal for the life that existed, and if the life-forms occasionally spilled over into those neighboring environments, then natural selection would start selecting variants from the original population which were more able to survive and thrive in that slightly different, and as yet unoccupied, environment. Important environmental variations undoubtedly existed in terms of temperature, salinity, concentration of any of several dissolved "nutrients," depth, light, and pressure. If mutations occurred (and they apparently did) that introduced variation into the first population of organisms, and if neighboring and nearby disjunct environments varied slightly in "quality of life" factors, then it is inevitable that selection for diversification would have taken off, that is—adaptive radiation would have begun. Diversification or speciation of life-forms eventually leads to some level of competition between those species, which itself leads to enhanced differences between species that come to coexist in the environment or mingle slightly along environmental boundaries (as in the ecological concept of "character displacement").

In short, if the environment varies either along gradients or across distinct junctions, evolving populations will eventually radiate to adapt to all the potential "niches" available, as fruit flies have seemingly done in Hawaii, as finches have done in the Galapagos, and as marsupials have done in Australia. As the mathematician character in the film *Jurassic Park* said: "life will find a way." Once there are reproducing organisms with exponential reproductive potential, and genes that occasionally mutate to provide heritable variation, natural selection and adaptation are the logical outcomes. Adapting evolution typically insures that any available space and resources will not go to waste for long. This adaptation and diversification into new and different environments would often require some increase in complexity as in the addition of a new enzyme, a new structural protein, a new organelle, a new membrane channel, a new or altered metabolic pathway, a new protective outer covering, a new symbiotic union, a new behavior, or a new body plan.

In the previous chapter, the twelfth listed characteristic of life involved the low level of entropy present in, and characteristic of, all living organisms. This structural complexity (the opposite of high entropy) is easily seen in the several recognized levels of biological complexity or structure, often covered in introductory biology texts. Both living and nonliving systems contain subatomic particles, atoms, and molecules, but in the nonliving realm, molecules are never as large and complex as those found in living systems. The highest level of molecular structure to be found in the nonliving realm would be that found in naturally occurring crystals, where at least the bonding arrangements of the molecules are highly aligned and structured.

Only in the living realm are there any regularly occurring structural levels that go beyond crystals in terms of ordered and complex structure. The *cellular level* is an obvious level of structure far above that of molecules, and some sources would insert the *organelle level* between molecules and cells in the hierarchy of life's complexity levels. Next after cellular comes of course the *organism level*. This level poses a slight problem because many organisms are single cells, and in that case these two levels coincide. It is well to remember here that some unicellular organisms are the most complex and structured cells known to exist. Because they must carry out all of life's many processes within the confines of a single cell, unicells—particularly the eukaryotic protists, are often much more complex than any of the specialized cells of a multicellular organism, since specialized cells have fewer jobs to carry out.

Some sources would insert a *tissue level* and an *organ level* between the cellular and organism levels, but that would of course only apply to multicellular organisms, and many of those don't have what could really be thought of as true tissues or organs. After the organism level comes the *population level*, since all organisms are part of a natural population. Do populations actually have structure in the form of predictable population-level interactions and phenomena? Well, do schools of fish have spatial structure? Do wolf packs have dominance structure? Do ants have caste structure? Of course the answer is yes in these and millions of other species where at least some type of family-level to population-level phenomena are regularly observed. Any good ecology book would mention several other examples of population-level structure(s), and a large subdiscipline of ecology is termed "population ecology."

Some sources would next list the species level, but that is really more of a conceptual grouping than a level of structural order and complexity. Populations interact with other populations in an ecological *community level* of structure. Again, do communities really have structure? Well, community ecologists certainly think so, as in—that's what they study and try to decipher—the myriad structured and regular interactions between populations of organisms such as plants and pollinators, parasites and hosts, predators and prey, grazers and plants, etc. The collective results of these regular and recurring interactions are what we sometimes refer to as "the balance of nature." Though this idea of a balance in nature is not well defined or agreed upon by ecologists, and has even fallen out of favor in some respects, there are certainly some regulating factors and interactions that allow communities to persist and to have "structure."

Beyond the community level we traditionally list the *ecosystem* and the *ecosphere* (biosphere) as the final two levels in the living realm. The ecosystem includes the living community members and their interactions along with the nonliving or abiotic components with which the living members must interact. Abiotic factors include temperature, wind, salinity, altitude, pressure, light, and others. Adding in the abiotic interactions only increases the number and complexity of regular interactions that occur within the community. The famous (though highly questionable) Gaia Hypothesis proposes that the whole ecosphere interacts in precise and purposeful ways which stabilize the biosphere (proposing yet more complexity and structure!). In short, life is complex, and it is complex at multiple levels that go far beyond anything found in the nonliving realm. In fact a single DNA molecule could potentially contain more structural complexity than a whole spiral galaxy of suns and solar systems if it could be known that no life existed in that galaxy.

There are so many amazing aspects concerning complexity in our living world. It could certainly be argued that one major component of biodiversity is the range of complexity across which it exists. The simplest prokaryote (bacteria) is at least thousands of times less complex than a human or any other complex animal. Even within the animal kingdom there exists a truly vast range of complexity in body plans. Sponges typically have only six to eight types of specialized cells and no organs or systems of any kind. Mammals are composed of roughly 30-plus-cell categories (many with subcategories), several organs and organ systems, some of which are unbelievably complex (such as the immune or nervous systems). The smallest animals in terms of cell numbers are small parasites of squid and other cephalopods called Rhombozoans. These animals in their parasitic stage are composed of only 20–30 cells total (and only three to four cells types). By contrast, a human contains on the order of 60 trillion cells (printed estimates vary widely, though all are in the trillions). Since we can't begin to comprehend the number for humans, let's not even bother with how many cells a whale might contain!

Why all this complexity and diversity? As mentioned earlier, sometimes the process of adapting to a new environment requires a change, something new (or *derived* in evolutionary terms) that will allow greater success in the new or changed environment. Below is a list of factors which have each contributed in some way to both complexity and diversity in the living world. This list is likely incomplete, but hopefully it will aid in understanding some of the driving forces in the buildup of both complexity and diversity. Of course natural selection is almost always the primary "force" involved in retaining any particular increase in diversity or complexity in the living realm. "Retaining" is the correct term here because natural selection doesn't create the derived features of life rather it eliminates those derived features which don't work. It is much more a negative process than it is a positive process (see the Mayr quote at the start of this chapter). For some, this point is one of the hardest to comprehend and incorporate into one's understanding of evolution by natural selection. Almost nothing in the long history of life was "selected for," rather those new or old features which were less efficient in contributing to biological selfishness were removed from life's playing field by natural selection.

FACTORS CONTRIBUTING TO INCREASES IN COMPLEXITY AND/OR BIODIVERSITY

You will note that some of the items listed below overlap to some degree, but this is simply unavoidable. Also, there certainly is no "best" order for listing these items, so the order here is not meant to imply relative importance.

1. *Speciation:* Most obviously contributes to biodiversity at the species level, and typically at several other levels as well (genetic, behavioral, morphological, physiological, etc.).
2. *Competition:* Though it can potentially lead to extinction, competition can also drive character displacement and environmental

partitioning, which can deepen the level of biodiversity between competing populations, species, etc.

3. *Extinction:* Yes, extinction has contributed to diversity—to both temporal diversity (Chapter 8), and to "deep" diversity (Chapter 5). Extinction also creates open niches and new opportunities for remaining species, which can then adapt and speciate in new and unique ways. Without the last mass extinction ending the Cretaceous, mammals would likely never have become so successful and diverse, though dinosaurs and other reptile groups might have become more so.

4. *Mutation:* The origin of new genes and gene products that can increase biodiversity at many levels (molecular, cellular, morphological, metabolic, behavioral, etc.).

5. *Changing environments:* Disturbances, mass extinction events, continental drift, climatic changes, introduction of new species (competitors, predators, parasites), etc., all of which move adaptation in new directions in affected species. Events such as continental drift, mountain chain uplift, sea-level change, etc. can bring about allopatric conditions (physical separation) between populations of the same species and lead potentially to true speciation, thus obviously contributing to biodiversity.

   Ecosystem disturbances such as fires, storms, epidemic-level disease, etc. can disrupt climax communities in some areas and allow several pioneer or serial-stage species to coexist alongside climax or near-climax species in neighboring undisturbed areas. In ecology this "intermediate disturbance hypothesis" suggests that if disturbances are not too widespread and severe, they can contribute to overall species diversity by creating a patchwork landscape (or seascape) of communities in various stages of ecological succession, each containing species not typically found in neighboring areas. Also, disturbances can hold population levels of competing species below the environmental carrying capacity such that competitive exclusion of one species by the other will be postponed or avoided (allowing both to coexist).

6. *Sex and diploidy:* Obviously both of these "inventions" increase the potential variation (diversity) of offspring, which can increase both the speed of, and the chances for, successful adaptation to changing environments—adapting into new available niches, speciation, and avoidance of extinction.

7. *Gene duplications:* Some chromosomal mutations, diploidy, polyploidy, viral-mediated gene transfers, "jumping genes," etc., which allowed two or more copies of a gene to exist in each cell. These duplications allow one gene copy to carry out its normal and

necessary role in the cell, while the duplicate may be free to incorporate mutations and perhaps acquire a new role in the life of the cell/species. This process undoubtedly explains the family of hemoglobin molecules, the several variant glucose transporter molecules (discussed in Chapter 6), and many others.

8. *Multicellularity:* Whatever the reasons behind the multiple and independent origins of multicellularity in plants, animals, fungi, brown algae, red algae, etc., multicellularity in each case set the stage for radiating increases in complexity and diversity. The multicellular clades all gave rise to broad and divergent ranges of organismal structure and complexity at the cellular and tissue levels. Animals, which added in organ-level and behavioral complexity, are the most diverse (in terms of *known* species) of the five traditional kingdoms.

9. *Symbiosis:* Endosymbiotic contributions to eukaryotic cell origins, lichen symbioses and diversification, coral and zooxanthellae, termites and gut protists, leaf-cutter ants and their fungi, hydrothermal sea vent animals and chemosynthetic prokaryotes, etc. The list here is extremely long. Lynn Margulis is on track in stating that many increases in diversity have resulted from, or at least involved, the evolution of close symbiotic relationships. Since parasitism is a subcategory of symbiosis, and since most species are believed to be parasitic, this conclusion is near inescapable.

10. *Arms races:* Predator/prey or parasite/host escalations in adaptive countermeasures which can add to the complexity level of the respective species (as in host immune mechanisms and parasite countermeasures) and deepen the divergence of related species "caught" in separate arms races, such as two related predators specializing on different prey or two related parasites specializing on different hosts.

11. *Addition of characters:* Obvious! Origin of the more complex eukaryotic cells, of multicellularity, of new metabolic pathways, of new structures such as hair in mammals, feathers in birds, wings in most insects, flowers in flowering plants, flagella in some bacteria, and countless others. The addition of a new character is often the result of what is termed an "exaptation." Exaptations are new structures that were derived from old structures of a different form and/or function. In fish, one early group of lungfish that had lived in shallow freshwaters diversified and adapted to life in deeper waters, including the oceans. Their lungs were no longer needed to supplement respiration because large deep bodies of water always contain plenty of dissolved oxygen. One of the two lungs was lost altogether, but the remaining lung was modified into a structure

called a swim bladder which took on the new function of allowing the fish to achieve neutral buoyancy at various depths. Cilia in our protist ancestors were used almost exclusively for movement through an aquatic environment, but cilia have been exapted in some modern descendents like ourselves for aiding in egg movement through oviducts and for moving mucus up our respiratory tree—thus helping to clean and maintain the lungs. Here the exaptation was one of function rather than of form. There are countless examples of exaptations including even certain genes that have taken on new and diverse functions in various lines of descent.

12. *Loss or reduction of characters:* Much biodiversity actually arises from the loss or near loss of characters in various groups and species. We lost our tail and body hair (almost) making us quite distinct from most other primates, snakes lost legs and one lung, making them distinct from lizards, many parasitic worms lost their sense of photoreception, cave animals often lose their sight and skin pigments, tapeworms lost their entire digestive tract, some parasitic plants lost their chlorophyll and photosynthetic pathways, while some animals became anaerobic and lost metabolic pathways like Krebs Cycle and the Electron Transport Chain.

13. *Sexual selection:* Intrasexual selection—as when males fight or compete with other males for access to females and/or intersexual selection (mate choice)—where one sex (typically the female) is especially choosy about selecting certain mate characteristics. Both of these forms of sexual selection have been the driving force behind many of the divergent differences between males and females of various animal species (sexual dimorphism). They have also worked to deepen or escalate the differences between species. Sexual selection is not separate from natural selection (as some writings suggest), rather it is a subcategory of natural selection, and a very potent one in some species.

14. *Learning in animals:* Simpler animals have little in the way of learning abilities, especially long-term learning. More complex animals often can learn things that alter their behavior and their lives (temporally). Some animals even have learned cultures which can vary from one population to another within the same species (as they most obviously do in our species), and which can change and diversify at rates that far exceed those of natural selection (more in Chapter 14).

15. *Social behavior in animals:* Many animal species display social behavior to some degree, and in many of those, social behavior has been

a springboard toward division of labor and even morphological diversification of group members. The social insects with their various castes display this aspect of diversification and social complexity in the extreme. Those insect groups that have evolved these complexities are among the most successful and diverse of animal species (consider the ants alone at over 9,000 species).

16. *Biodiversity:* Biodiversity itself is a cause of biodiversity since it creates more potential niches: hosts, food/prey, habitat (coral reefs and rain forest canopies), competitors (which can drive character displacement), etc. Keystone species are defined as those species on which the survival of many other species depends, and their presence obviously contributes to biodiversity at the community/ ecosystem level (corals in a coral reef, kelp in a kelp "forest," or pollinators of a dominant plant species). A somewhat common phenomenon among parasitic species is speciation in conjunction with (or following on the heels of) host speciation (a particularly clear case of biodiversity driving biodiversity). A diversity of infectious agents and parasites in our evolutionary history also resulted in a diversity of immune mechanisms, each specialized for dealing with different groups of invaders.

17. *Emergent properties:* When organisms become especially complex, another source of both complexity and diversity comes into play, that being the phenomena of *emergent properties.* Emergent properties are those aspects of organisms (or other complex systems) that are not predictable or expected based on knowledge of the "parts" making up the whole organism. Though the reductionist approach has been extremely productive in biology, as in the other sciences, it turns out that not all phenomena can be predicted or explained as the "sum" of their interacting parts. Much of biodiversity exists as emergent properties in whole organisms, properties that are not even suspected from studying the organism's components.

Complex behaviors are especially good examples of emergent properties. It is true that some very simple behaviors such as reflexes can and have been sufficiently explained as the interaction of a few identified nerves and muscles. Most of the more complex behaviors however involve so many neural circuits and so much complex processing that we haven't a clue how such behaviors come about, except to say that they involve some very complex neural interactions. No amount of knowledge of neural anatomy along with precise neural monitoring would predict that Monarch butterflies innately migrate thousands of miles in a huge North American circuit. How could one ever find, in the tangle of neurons and their interconnections, the "map," the list of cues, the triggers for initiating and stopping flight, and whatever else the Monarchs use to find their way? Although such information must be encoded in the neural anatomy/chemistry of the Monarch, it is a foreign language which we have no power to translate *from the bottom up*. In theory (and likely only in theory) the mechanisms might be elucidated from the top down (monitoring the precise neural firings associated with each step of the migration), but little if any progress could be made from the bottom up (the reductionist approach). Even though a butterfly's nervous system is far simpler than a mammal's, it is too complex for us to even begin to discover its more complex processing. Such emergent behaviors are a huge part of biodiversity, and most animal behavior involves far more than simple reflexes. We will delve deeper into behavioral diversity in Chapter 14, but for now the point is that many behaviors and other emergent properties contribute significantly to the complexity and diversity of some groups of organisms.

Getting back to why evolution has generated so much diversity, there is one major reason that often goes overlooked, or at least goes unmentioned. This reason has to do with the fundamental workings of life and natural selection. Unlike the older idea that life was a progression of forms toward higher complexity and "perfection," along something previously called the ladder of life, biologists now understand that life clearly has no ultimate goals of form, complexity, or intelligence. Complexity and intelligence may have been attained by a few branches of the radiating tree of life, but only as rare isolated variants in the larger picture where much or most of life's diversity is still at the level of prokaryotic cells (bacteria and their kin) and the more complex unicellular species. We are only now starting to understand this point, which was the main topic of *Full House* (1996) by Stephen Jay Gould, one of his better books if you don't mind his long and supposedly related digression into baseball statistics.

If there are no preset guiding principles, or plans, or set steps to constrain the four billion years of life's evolution, and if environments have continued to change—thus forcing and allowing more diversification through natural selection, then there is little wonder that many millions of diverse species have arisen throughout this immense span of time. Only a couple of hundred years ago many scientists still accepted the ladder of life concept and believed that all of the "lower" forms of life were predictably evolving their way up to become the "higher" forms of life. If this were indeed true, then why is there so much diversity at the same level? Why are there over 8,000 species of polychaete worms, none of which is noticeably more "advanced" in its structure or behavior than the others? Why are there 10,000 species of millipedes (twice the number of mammal species) which all have the same basic body plan, one that hasn't varied significantly in 400,000,000 years? The great diversity at any one level of structure and complexity should have been an obvious challenge to the progressionist notions of earlier biologists, not to mention the diversity of whole extinct groups such as the trilobites, which never evolved into any so-called higher forms. Of course these earlier workers were not as aware of the history and immensity of biodiversity as we are today, but they should have been aware of enough diversity to make it questionable as to how all those life-forms could be correlated or assigned to particular rungs on a presumed ladder of life.

The point is that life is diverse because it is relatively free to be so. Life is not constrained (so far as science can tell) by any divine or mystical plan, direction, or set of steps. The contingencies of environmental change, natural selection, and random events (such as massive asteroids colliding with the planet) have therefore worked to generate an unbelievably diverse assemblage of species throughout life's long history.

Another old idea that still holds sway in some writings is the idea that every species in nature plays an important, if not essential, role in its respective ecosystem. This idea, if true, might suggest that there is some set number of roles that each ecosystem "should have," each role-player contributing in some way to the efficient running of that ecosystem. This idea too is unsupported and easily disproved by the numerous changes in flora and fauna which we know to have occurred without any major noticeable effect in the total ecosystem (except of course for those individual losses and additions).

In ecology, we learn of symbiotic relationships where two species interact in one of three basic ways: mutualism, commensalism, or parasitism. Though commensalism seems to be the least common of the three categories, there are still numerous examples of commensalism in nature. There are thousands of species of bacteria and protists that live as commensals in the digestive tracts of vertebrates (along with many that are definitely mutualists). Though a few of these may be in some small immeasurable way affecting the host animal, most seem to have no positive or negative effect on their host and feed mainly on one another, unabsorbed remains of digestion, or on sloughed epithelial cells, doing no measurable harm to the host. These commensals are often found nowhere else in nature except in the dark recesses of their host guts (and as dormant cysts in fecal waste—waiting to be ingested by another host). Do these commensals play some important role in the ecosystems of which they are undeniably a part? It seems the only logical conclusion one could make is that they do not. Both their hosts and ecosystems are essentially unaffected by either their presence or their absence. Such organisms are certainly a part of nature and a component of biodiversity, but they are not filling some preset or required role in the environment. They exist because they can exist. They have diversified into a niche which had "ecological room" for them, and they have thrived.

There are no constrains requiring species to have purposes. Commensals *can* exist, therefore many have evolved, and in the process they have raised the level of biodiversity even higher. Populations are indeed relatively free to diversify into countless niches, in some cases recreating those niches as they evolve into them (as we have certainly done). Species can add to biodiversity by becoming either more complex or less complex (something the progressionists didn't understand or appreciate). Groups can move from one environment into another and even back again (as whales and sea snakes have moved back into the seas). This freedom from constraints is unquestionably one of the fundamental reasons for the immensity of biodiversity on our planet. The belief that all species have essential ecological roles is rooted in part in religious metaphysics, but may also be an example of extreme adaptationist thinking, something warned against by Stephen Jay Gould and other modern evolutionists.

We now have fairly good evidence that most of the DNA in our cells plays no role whatsoever in coding for the structure and functioning of our complex bodies. The estimates for this useless DNA run to well over 90 percent, though this estimate is certainly premature since we really don't know just yet that all that so-called "junk DNA" is really useless and without some regulatory function. A major hypothesis concerning these extra DNA sequences is that they are molecular parasites of the cells and the species that contain them. They exist and are passed on because they *can* exist and be passed on. There's not much to stop parasitic DNA from being reproduced, even though it may pose some very small drain on the materials and energies of the cells required to duplicate this parasitic DNA. This useless DNA, parasitic or not, is another challenge to extreme adaptationist thinking which posits a role for every molecule, structure, species, etc. known to have evolved. People outside of biology and science often have the same notion—that every species plays an important role in the workings of nature, but to quote Francis Bacon: "The human understanding on account of its own nature readily supposes a greater order and uniformity in things than it finds."

In short, there really are far fewer constraints on what can exist (in terms of living diversity at every level) than even some biologists would suppose. Let me make it perfectly clear that I value every bit of living diversity whether it is known to play an important role in its environment or not. If we are to value anything in this world, I would argue that we must value life in all its various forms and levels *simply because it exists.* As Daniel Dennett wrote in speaking of "The Tree of Life": "This world is sacred." Obviously, not all people feel this way, and not even all scientists feel this way, but many do. Dennett doesn't mean sacred in a religious sense, as he clearly isn't of the religious persuasion. If we understand and accept that we as a species are but one among millions of the contingent results of an unbelievably long and complex history of something called "life," many would argue that we must hold the whole process and its results in awe if we are to claim any shred of human aesthetic feeling.

In a somewhat strange little book by Alexander F. Skutch entitled *Harmony and Conflict in the Living World* (2000), Skutch (a well-known ornithologist and specialist on tropical birds) expounded some ideas and beliefs that are unorthodox to say the least. He suggests that nature "seeks" harmony in its interactions. He even suggests that we should promote harmony in nature by controlling biodiversity, that is, by aiding those species which contribute to the harmony of ecosystems and controlling or eliminating those that don't. He suggests that as much as 50 percent of species biodiversity contributes to "suffering and disharmony" in nature.

If this is what Skutch actually believes, he has far greater confidence in humans to understand nature and to make wise decisions than do the great majority of scientists. Also, he apparently believes we have the *right* to manipulate nature as we see fit, according to our own definition of harmony. It is true that we already manipulate nature on a massive scale, but not with the goal of making it more harmonious! We can at least understand some of the economic reasons for damaging ecosystems, but it is hard to imagine the altering of nature on a massive scale for the reason of making it "better."

Life almost always works best when it is left alone to do what it does best, and that is to evolve and adapt to whatever openings are available and to struggle for selfish survival and genetic fitness within the rich context of millions of other species. It could be easily argued that even the disease organisms and parasites of an ecosystem are of value and contribute to the "health" of those ecosystems. If we manipulated nature in a significant way, we couldn't very well call it "Nature," now could we? Most scientists, and especially most biologists, would agree that nature is worthy of our awe and appreciation (at least aesthetically), and if anything, deserves to simply be what it is.

Finally I must clarify that I did not mean to imply earlier that nature contains no constraints whatsoever on the evolution of biodiversity. Every biologist knows that nature is to some extent a constrained system due to:

- Constraints of available resources: Cave salamanders can't be very large—too little food to support large body size.

- Constraints of the energy pyramid: Animals can't specialize as twentieth-level consumers—due to the average 80–90 percent loss of energy with each consumer step.

- Constraints of physical laws: Grasshoppers can't grow to the size of sheep—their support, circulatory, and respiratory systems won't scale up efficiently enough to support such a size increase.

- Constraints of genetic variation: Many species went extinct because they did not contain sufficient variation with which to adapt to environmental change, while others which *might* have speciated under certain conditions failed to do so for lack of sufficient genetic variation.

- Constraints of phylogenetic position: Bacteria can evolve faster than mice, and mice can evolve faster than elephants—due to the differing generation times and the number of offspring produced (faster generation times plus greater numbers of offspring equals potential for rapid evolution and speciation). Also, evolution can only tinker with what is already present. Wing evolution (changes to existing wings) can occur in birds and insects, but not in earthworms or humans.

- Constraints of available niches: You can't specialize as a symbiont of tapeworm guts because tapeworms don't have guts—intestines that is.

- Constraints of competition: Competition was listed earlier as one of the factors which enhances diversity, so how can it also be a constraining factor? Paradoxically, it is both. Severe and prolonged competition between two species will not uncommonly lead to the extinction of one. Niches that are already filled with well-adapted species with likely not be available to fragmented groups from other species (potential seeds of speciation) because the resident species will easily outcompete the new arrival. Competition is a part of natural selection, which itself is a constraining factor on diversity but at the same time is a prime creator of biodiversity. Remembering that natural selection is in essence a selective killer of the maladaptive, it is truly awesome to consider the expanse of life's complexity and diversity that has arisen from a drama characterized by death on a scale of such magnitude.

There are then several constraints or limiting factors on total diversity in addition to those of common descent and life's properties discussed in Chapter 1, nevertheless—life has obviously flourished and diversified greatly within these constraints.

3

Species Diversity

Our museums are glutted with new species. We don't have time to describe
more than a small fraction of those pouring in each year.
—E. O. Wilson, *The Diversity of Life*, 1992

And once new species have been described and placed in a museum, what
do we really know about them?
—Howard Ensign Evans, *Life on a Little-Known Planet*, 1968

How many species of organisms are currently living on our planet? This
sounds like such a simple question, one that many people would guess
to have a known answer—all that is needed is to look it up in some sort
of biology book. Unfortunately, like a lot of simple-sounding questions
in biology, this one does not have a simple answer. In truth, it has no
available answer at all. But even though there is no one central agency
responsible for keeping track of the current total of known species, we
are relatively certain that there are around 2,000,000 *known* species, but
the known (or described) number is of course not the same thing as
the *actual* number of living species. It now seems clear to those who
are at the forefront of this area of biological knowledge that at best
we only know something like 25 percent of the earth's actual species.
Estimates of actual extant species range from a low of 5,000,000 to a
high of 100,000,000 and more. Sadly, some of those 2 million known
species have likely become extinct since they were described.

Some have suggested that if we were to be questioned by intelligent
aliens about our world, one topic they would surely be curious about
would be the diversity of life-forms on our planet. Sadly, most members
of our supposedly intelligent species would be almost clueless as to
describing the diversity of life on our planet, and would surely not be able

to come up with anything like a knowledgeable number of species, unless by chance. Most people could not even convey an adequate definition of the terms "life" and "species," so unless these aliens happened to abduct a biologist or a well-read layperson, they would be largely out of luck in obtaining answers about the life of our planet.

How many is 2,000,000? We can easily say the number, and many people have over $2,000,000 of net worth. But can we really comprehend this number? If we were to view a slide of each species for only 5 seconds, for 8 hours a day (that's 5,760 slides/day!), it would take 347.2 eight-hour days to view the known life on Earth. Even though the time taken is less than a year, which we can comprehend, the number of slides (species) viewed in this example would quickly push the limits of true comprehension well within the first week, remembering that each of those species is unique and different in at least some ways from all the rest.

In 1758, in the tenth edition of his *Systema Naturae*, Linnaeus listed around 9,000 species of plants and animals. We have come a long way since then, but our efforts are still relatively in their infancy if at least 75 percent of earth's species await discovery. It's true that much of the globe has been explored, but to say that a tropical river basin has been explored usually only means that it has been mapped, with a few photos and drawings made, a few specimens taken, and perhaps a botanist and a zoologist were along to document what they could on a few excursions inland from the river. They would likely not have covered more than a tiny fraction of the square miles in that basin, nor would they have had the expertise to identify, or even collect, most of the biodiversity present in that basin. Most life is small and lives in isolated microhabitats such as in small tributaries and seeps, in the soil, in the humus of the forest floor, in and on the bark of trees, in the "guts" of the animals, in the treetops some 100 feet up, etc. You would need a team of several biological specialists and at least a year of collection and study to even begin to do justice to the diversity of any tropical river basin.

The oceans which cover 71 percent of our planet are not well explored either, and they contain many undiscovered species from every kingdom except perhaps the plants. The deep ocean basins are only now beginning to be explored on a regular and systematic basis, and newly investigated areas almost never fail to reveal several species never before seen or described. Much of the ocean's undiscovered species diversity likely exists among the microbial communities: the prokaryotes, fungi, and protists.

Before continuing, it seems appropriate to address the question of exactly what is meant by this term "species." Biologists often have a more specific definition of the term than laypeople, but even biologists can differ on exactly what they take a species to be. It turns out that there are at least a handful of regularly used species concepts in biology, so before we try and ask how many species exist, we should explore briefly these various concepts.

THE MORPHOLOGICAL SPECIES CONCEPT

Morphology refers to the anatomy and structure of an organism. Though most biologists immediately think of the large-scale structures, which are most easily visible such as color, shape, size, texture, number of appendages, presence or absence of hair, etc., internal and even microscopic morphology is often essential for the identification of some species (obviously so for protists). Naturalists and environmental biologists are very familiar with taxonomic "keys" which allow one to decipher which species of tree, or insect, or snake one is observing. Through a series of steps, each involving a choice between two morphological descriptions, one can, with patience, arrive at the correct species name for the organism in question (assuming it is not a new species not yet included in the key). Most early attempts to classify organisms utilized the morphological species concept to a great extent. Aristotle and Linnaeus certainly classified mainly by morphology. Long-extinct organisms can only by classified by the morphologies detectable in their fossils.

Biologists wish that all species were easily identifiable through obvious morphological differences, but unfortunately that is not the case. Many organisms go through complex life cycles in which two or more distinct morphologies are present. A larval mosquito looks nothing like the adult, and a larval sea star (starfish) likewise looks nothing like the adult. Why should we only emphasize the adult when in some cases it is the briefest stage in the life cycle? In the case of mayflies, the larvae live and grow for several months in aquatic environments where they are important components of the aquatic ecosystem. When they emerge as adult terrestrial insects, they live only for a matter of hours during which their only role is to mate and insure the next generation of mayflies. Though the adults are eaten by many animals, their short existence and single objective (reproduction) argue that the aquatic larvae are the really important stage in this life cycle. In fact, there are separate keys for both adult and larval mayflies, since both are important to aquatic ecologists.

THE BIOLOGICAL SPECIES CONCEPT

Most biology texts stress this concept as "the" definition of what a species is, though it applies only to sexually reproducing species. It states that a species consists of a population or populations of individuals which can potentially interbreed successfully under natural conditions if given the opportunity. In short, if a male successfully mates (or could mate) with a female in nature, then they belong to the same species. If two males could potentially mate with the same female successfully, then they are of the same species, etc. Successfully here means that the mating will produce offspring which will be as capable of interbreeding with other individuals in the species as were their parents. This rules out most hybridizations between species (which do rarely occur), because most hybrid offspring are not as viable in health or reproductive potential as the offspring produced within single species.

The part of the definition stating "under natural conditions" is required because many species will hybridize in zoos or on farms that would not do so in nature. Domestication often removes much of natural behavior (especially in mammals) such that lions will hybridize with tigers to produce ligers, donkeys will hybridize with horses to produce mules, gibbons will hybridize with siamangs to produce siabons, etc. Often in nature the ranges of these species don't overlap, but even where they do, they do not typically hybridize when they live as part of natural populations.

This definition can easily cover populations of monoecious individuals which, though being both male and female, must usually mate with another individual for fertilization to take place (in both individuals). It can also be applied to organisms having complex life cycles which include both asexual and sexual reproduction. Even if most reproduction is accomplished asexually, when sexual reproduction is carried out—it must typically occur between individuals of the same species.

Many writers have expressed the idea that because of this relatively unambiguous definition, the species level (for sexually reproducing organisms) is the only rank level in our classification which is "real" and consistent from birds, to fish, to pine trees, to fungi, to *Paramecium*, etc. Though other taxonomic ranks (genera, families, orders, etc.) typically represent clades, there is no consistent way of recognizing where to draw the lines between two families that would equate from snails to ferns to ciliates.

Certainly if one looks at all the variety in nature, one can find instances which only partially fit the biological species concept and are thus problematic. Almost every general statement or definition in biology has its exceptions, but we still find value and usefulness in those general statements and definitions. For many of the earth's species, the biological species concept has been, and will likely continue to be the best and most meaningful concept available.

THE EVOLUTIONARY (PHYLOGENETIC) SPECIES CONCEPT

According to this concept, species simply represent the distinct "lines" which are represented in cladograms. Where one line continues forward along the time axis, one species is represented. Where one line splits into two, one species has become two. This concept emphasizes the branching species lineages that can (at least in theory) be represented along a time line to show their respective longevity, their origins, their phylogenetic connections, and thus their degree of relationship to other lineages, etc. This concept in short emphasizes the historical component of species, both living and extinct. This concept can comfortably include asexually reproducing organisms which the biological species concept cannot, but it is not very pragmatically useful to a working biologist. This concept certainly has its strengths and drawbacks, as do all the other species concepts.

ECOLOGICAL SPECIES CONCEPT

This concept in essence derives from the idea that each species occupies a unique multidimensional ecological niche. No two species could possibly occupy exactly the same niche when all parameters are included such as geographical range, temperature range, moisture range and tolerances, symbionts, diet, etc. There are other side aspects and issues associated with this concept, but since it is highly impractical (if not impossible) to define and measure all the naturally occurring parameters of *any* species niche, this concept is largely theoretical with little utilitarian value.

MOLECULAR SPECIES CONCEPT

This concept is now being developed especially in regard to asexually reproducing organisms like most of the prokaryotes and protists. The recently acquired ability to directly compare the similarity of DNA, RNA, proteins, etc. between and among "groups," and thus determine the degree or percent similarity of those groups, is the basis of this concept. If two pure bacterial samples vary by some significant percent of their DNA or RNA, then they are recognized as distinct species. The complexities of which genes, which RNA molecules, which proteins, and what thresholds of similarity are to be used in these comparisons are issues still being ironed out by the specialists in this field. Recent published reports on the progress in this area seem to indicate great utility and promise for getting a handle on species diversity and phylogeny. Though especially useful in asexual groups, this concept has some utility in the sexual groups as well.

Unfortunately, these five concepts are not the only ones being considered and argued in biology, and we will not mention the others. As you can see, there exists a "diversity" of ideas on what species diversity actually means. Biologists are obviously still searching for the "best" concept(s) of what we mean when we use the term "species." Each of the various concepts has its strengths and weaknesses, and there is certainly some potential agreement and overlap between some of those listed. Perhaps one "hybrid concept" will eventually result, or a few well-defined concepts will be accepted as uniquely applicable to different groups of organisms (such as asexual vs. sexual). The ongoing debate concerning the respective strengths and weaknesses of these various species concepts is undoubtedly a healthy enterprise, one which stimulates thinking and which hopefully leads to consensus.

For the past 30 years or so, biologists have divided life into five kingdoms based on major distinctions of morphology and metabolism, but some workers have recently come out with the idea that five is far too few (see Chapter 15). Their ideas suggest that the current kingdoms Monera and Protoctista should each be divided into several Kingdoms reflecting the deep genetic and metabolic differences in their diverse and truly ancient phylogenetic lines. The prokaryotic *Kingdom Monera*, now recognized as containing two of the three "Domains" of living organisms (Bacteria and Archaea), currently holds just over 5,000 described species of prokaryotes (organisms lacking nuclei in their cells). Estimates of the actual number of prokaryotes start at 50,000 and run into the hundreds of millions! The definition of the term "species" for prokaryotes is now being based on the percent of DNA variance between "strains" (molecular species concept), rather than on potential for interbreeding as in most groups of eukaryotes. The earth is still relatively unexplored in terms of prokaryotes because only recently have workers realized that most microbes are not easily collected, and most are not easily cultured in a lab environment. Basic exploration of the planet in terms of microbial biology is only now really beginning, so expect that 5,000 to increase rapidly and dramatically in the near future.

In the *Kingdom Protista or Protoctista*, estimates of known species range up to 250,000. As mentioned earlier, this kingdom too may end up being split into several kingdoms in the near future. As with the Monera, species distinctions are hard to base on the concept of interbreeding populations, since many protists reproduce only by asexual methods. Again, the use of DNA and other molecular comparisons and analysis will likely become the standard tools for discovering and defining species diversity in this huge assemblage. Morphology is often not an ideal clue to species recognition in one-celled organisms. There are several species of *Paramecium* (proven through breeding experiments; *Paramecium* do conjugate) which are morphologically indistinguishable, that is, there are likely large numbers of sibling species within many of the protist groups.

The *kingdom of the Fungi* holds roughly 70,000 identified species, with estimates of actual species ranging up to 1,500,000. Fungi are eukaryotic, mostly multicellular, and have heterotrophic absorptive nutrition. Fungi typically consist of threadlike hyphae which are syncytial. The hyphae have cell walls which typically contain chitin (a complex polysaccharide). Fungi form haploid reproductive cells called spores through meiosis. There are almost certainly many thousands of undescribed fungal species in the soils, forests, and aquatic environments of our world. Many fungi are parasitic on animals and plants, and there must logically be hordes of undiscovered species of these as well.

The *Plant kingdom* currently contains around 250,000 described species. Plants are eukaryotic, mostly or all multicellular (depending on whether the green algae are included), and use photosynthetic autotrophy (except for a few parasitic plants which are parasites on other plants). Plant cells also have cell walls, typically made of cellulose and lignin, and in their sexual reproduction plants develop from multicellular embryos. Though the discovery of new plant species continues today at a rapid pace, this kingdom is most likely the best known in terms of its species diversity. Plants are mostly terrestrial, mostly obvious exposed macroscopic organisms, and they cannot run and hide at the approach of a field botanist exploring the plant community. Also, there are disproportionately more taxonomists who specialize in plants as opposed to any of the other four traditional kingdoms. For all these reasons plants are relatively well known and characterized, while we still have a long way to go in elucidating species diversity in the other Kingdoms.

The *Animal Kingdom* is by far the largest of the kingdoms in terms of described species, standing at around 1,300,000 species. Animals are eukaryotic and multicellular. They are typically heterotrophic with ingestive (holozoic) nutrition. Animals lack cell walls, but like plants they develop from multicellular embryos, though some also have asexual methods of reproduction. The animal kingdom continues to be a hotbed of species discovery with at least one new animal species being discovered every day or so. Most new discoveries are insects, worms, and other small and inconspicuous animals.

Why animals are so diverse is a question with many potential answers. The main reason why animals are so diverse is that insects are animals, and insects make up roughly 70–75 percent of all known animal species. Obviously insects are the most successful group of organisms (at least at the class level) that has ever existed. Most insects are small (can adapt to numerous microhabitats), produce many offspring and have short life spans (can evolve and adapt rapidly), and have wings for dispersal (can move into new and different environments easily). Most insects also undergo complete metamorphosis in which the larval young look drastically different from the adult (maggots—flies, etc.), live in different parts of the environment, and usually lack overlap with adults in terms of diet (reduces larval/adult competition). These and other features have undoubtedly contributed to the huge success of the insects, though even without insects the animal kingdom would still be extremely diverse in terms of the deeper diversity of classes and phyla.

No one can say precisely how many new species are discovered each day, week, or year, but such discoveries are occurring at an amazing rate. Between 1983 and 1993 over 450 new species of mammals were described. Between 1985 and 1995 over 750 new species of amphibians were described. Roughly 2000 new plant species were discovered through the 1900s. Keep in mind that each newly discovered species undoubtedly has several endemic symbionts (parasites, mutualists, and commensals) which will also be counted as new species as soon as someone takes the time and effort to document their existence. Many of the earth's species are undoubtedly sibling species that cannot be distinguished morphologically from other closely related species. Only through careful breeding studies or molecular comparisons will these species be discovered, named, and added to our inventory of "life on earth."

One last point worth making concerning those species which have been discovered and described is that very little is known about most of these species. Many species are known only from one or two specimens which were collected and described in a scientific paper. A morphological description, location, and date are all that are known for a large percent of our described species. In some cases, only the female or male is known and the other sex has never been seen or collected. For such species, even though they are "in" the scientific database, we know only that they exist and little more. We know nothing of the relative importance of that species in its ecosystem, its functional role, its behavior, its metabolism, its physiology, or its geographical range.

Each species holds a small "universe" of potential biological information. The biochemistry, physiology, morphology, behavior, and ecology of each species is at least in some ways unique, and worthy of study and comparison. Only by learning everything we can about each species can we approach a complete understanding of the life on our planet, a goal that is admittedly unrealistic and unattainable, but one which assures that the horizons of potential biological knowledge will continue to provide opportunity for new discovery and knowledge far into the future. E. O. Wilson commented that we now know all the genes of the human genome, yet we are still largely ignorant of the number and variety of life-forms we share the planet with. Wilson obviously finds this comparison ironic, as do most other biologists.

Certainly most people value knowledge that has pragmatic value, however most scientists believe that knowledge is good in and of itself, no matter its immediate utility. Perhaps it is a question of how we define utility. Lynn Margulis wrote: "Evolutionary science deserves to be much better understood." In the same vein most biologists would say that knowledge about *who* our fellow beings on this small planet are deserves to be better known and comprehended. There should be no connection or assumption that the species we have discovered are the "important ones," while those yet unknown hold little importance or value.

One of the most important groups of organisms on earth was discovered only in the late 1980s. Species of the cyanobacteria genus *Prochlorococcus* contained individuals so small that they escaped detection for many decades of marine research. Several species of *Prochlorococcus* have now been discovered. These photosynthetic prokaryotes constitute one of the most important sources of marine primary production through photosynthesis. In the oceans they form a major fraction of the base of food chains which support many species of consumers at higher trophic levels. They also soak up carbon dioxide from the atmosphere and thus play a role in regulating atmospheric CO_2 levels. Without them CO_2 levels would almost certainly rise faster and contribute to the already serious problem of global warming. Species such as those of

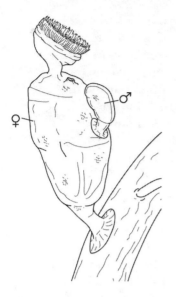

Figure 3.1 The odd little animal *Symbion pandora* (phylum Cycliophora), which lives commensally on the mouthparts of some marine lobsters. Note the dwarf male attached to the larger female. (*Figure by Jeff Dixon*)

Prochlorococcus which support and maintain several other species in their respective ecosystems are known as "keystone species." We certainly cannot understand, model, make predictions about, estimate the magnitude of perturbations to, conserve, or even appreciate an ecosystem fully without recognizing the role of its resident species.

Another recent (1995) discovery of great import was that of *Symbion pandora*, a species of tiny commensal animals (Figure 3.1) that were discovered residing on the mouthparts of Norwegian lobsters! The reason why these seemingly insignificant animals were an important find is that they were so different in their morphology and life cycle from any other known animals that they were judged to represent a new animal phylum (Cycliophora). Of course, the importance of this discovery is best appreciated by biologists, but it does significantly broaden our understanding of life on earth. The challenge now will be to figure out how Cycliophorans are related phylogenetically to the other 30-plus animal phyla.

Our growing understanding of life on earth continues to be an important frontier of knowledge where new discoveries are guaranteed. Who knows what importance each new species discovery will hold of either a pragmatic or scholarly nature? One thing is certain—we cannot approach a complete understanding of biodiversity and all its many parameters if we do not know the players (species) that make up that total biodiversity.

4

Diversity within Species

That no two individuals are alike is as true for the human population as of all other sexually reproducing organisms.

— Ernst Mayr, *This Is Biology*, 1997

When most biologists hear the term biodiversity, the first aspect of biodiversity that comes to their minds is species diversity, and rightly so. If asked about biodiversity within a species, some would mention only the genetic diversity within species, though of course genetic diversity exists at all levels from the individual (heterozygous loci) up to kingdoms and domains. Chapter 6 will cover genetic diversity in more detail both between species and within species. The purpose of the present chapter is to emphasize that there are many other parameters and aspects of biodiversity to be found within species other than genetic, though gene diversity is of course ultimately responsible for most of the following examples.

An awkward aspect of this chapter is that by necessity examples of *intraspecific* (within the same species) diversity occur in other chapters, and those chapters will be referenced later as we briefly focus more selectively on the topic of intraspecific diversity. Intraspecific diversity is a topic that typically goes unmentioned in many of the writings on biodiversity, so this brief chapter will attempt to direct attention to this important, though overlooked, aspect of biodiversity. This overlap of intraspecific diversity with other chapters and categories well illustrates the point that no matter how cleverly we attempt to categorize nature into discrete compartments, nature is far too subtle, graded, and repetitive to be easily subdivided and pigeonholed.

Back in Chapter 1, we covered 12 features common to living organisms. One of those universal properties of life is that all organisms are cellular. Cells are units of life, just as much as individuals are units of life. Cells certainly vary between species, and they vary greatly between species as unrelated as say kelp and pythons. But cells also show great diversity within species, especially so in multicellular species. Animals show this in the extreme, as anyone who has studied animal histology knows. Smooth muscles are easily distinguishable from skeletal muscles, as are either from epithelial cells, or nerve cells, or fat cells, etc. Our body cells are diploid with 46 chromosomes, while our sperm and egg cells are haploid with only 23 chromosomes. Our red blood cells are tiny, but some of our neurons are 2–3 feet in length. Mammals like ourselves are made up of over 30 distinct cell types, with finer divisions possible depending on how picky you are in classifying cells. Cells show diversity in size, shape, function, physiology, etc., and this is certainly one level of diversity that can and does exist within many of the earth's species.

In Chapter 6 the obvious point is made that in sexually reproducing species, essentially all the sexually produced offspring are genetically unique, which means that there exists a great diversity of genotypes and phenotypes within any one of these species. A most important point to be made here is that intraspecific genetic variation (diversity) at the level of individuals is typically required for speciation to occur. In short, species diversity is an outcome of the genetic diversity of individuals within species (coupled of course with natural selection). In some species, at least some genes and their phenotypes vary across subspecies or racial lines. The existence of subspecies is an indication of potential speciation. Race in humans is such a sensitive topic that some say it is wrong to even speak of races in humans. It all depends on how much genetic variation exists consistently between such groups, and knowledge in this area is still in its infancy. We probably need more information before we stamp out the idea of human races, no matter how politically correct such a stance might be. At any rate, there definitely are genetic and resulting phenotypic distinctions between some human groups. It is a near certainty that if an Eskimo couple conceives, and the female bears a child, that child would not be mistaken for a Caucasian, an Aborigine, or a Yanomamo Indian.

The real diversity within a species lies at the level of individuals. Humans are now believed to have between 25,000 and 30,000 genes, some of which of course have several variant alleles, but there are currently over 6,500,000,000 people on the earth, and except for a small percent of those who are identical twins, each human is genetically and phenotypically unique. Biodiversity among individuals in populous species would seem to be clearly the most diverse parameter of biodiversity. Though they may look superficially identical to human observers, consider that some species of marine zooplankton (and many other invertebrates) number conservatively in the hundreds of billions of individuals. Since most of these species utilize sexual reproduction, each of these individuals must be in some subtle ways unique from all the rest. Individual biodiversity (based on genetic diversity) and genetic biodiversity (considered separately) together must logically surpass all other parameters of biodiversity by a wide margin.

Chapter 7 deals in part with the great diversity of ecological interactions shown by the various species on earth. In this parameter too, intraspecific diversity is not uncommon. In many species of mosquito, it is only the female who engages in blood feeding. The males typically feed on plant nectar, if at all. This is also true for some species of horseflies, and even the morphology of the mouthparts may vary for these distinctly different and sex-specific ecological "roles." Ecological roles can change within complex life cycles. Mosquito larvae eat algae, while adults drink flower nectar or blood. It is hard not mention our own species in this respect. Eskimos have always been a mainly carnivorous people, taking their prey largely from the marine environment. Tropical hunter/gatherer peoples have a richer mix of food that is largely of plant origin and, depending on the particular group, often includes no foods of marine origin.

In Chapter 9 morphological diversity is addressed, again with some examples that occur within species. In humans, age-related changes of total size, body proportions, hair color and pattern, skin tone and texture, etc. add diversity within our species (and many others). In many species, males have morphologies quite distinct from females, and the reader is sure to think of several examples such as elephant seals, African lions, or orb-web spiders. For the many species that undergo metamorphosis, even greater diversity of morphology is present between successive stages in the cycle. The monarch caterpillar looks essentially nothing like the monarch butterfly it will become.

Larval sea stars likewise look nothing at all like the adult animal (Figure 4.1). Besides being much smaller, larval sea stars are pelagic/planktonic, swim by means of cilia, and are bilaterally symmetrical. The adult sea star is much larger, benthic (bottom dwelling), moves by a system of hydralic tube feet, and is radial in symmetry. It is hard to imagine more morphological diversity within a single life cycle. Some plants such as ferns go through changes of similar magnitude in their "alternation of generations." The large and familiar sporophyte plant,

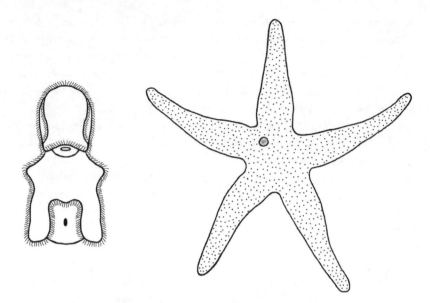

Figure 4.1 The extreme morphological contrast between the plank-tonic bipinnaria larva of the sea star (left) and the benthic adult sea star (right). Larva shown greatly enlarged. (*Figure by author*)

which we recognize as a fern, looks nothing like the tiny heart-shaped gametophyte plant of the same life cycle.

Chapter 9 briefly explains the three kinds of males that exist in Pacific salmon. Several fish species produce two or more male phenotypes. Often one is the large "normal" male typically associated with the species, while the other is smaller and also differs in behavior. Such males are often referred to as sneaker males. Such males often do not try to win females by building nests or performing costly courtship displays, rather they do what their name implies. They sneak around the periphery of a courting pair of their species, and when the moment is right and the female is releasing her eggs, the sneaker male darts in and adds his sperm to those of the larger male, often without even being noticed. Some percentage of the eggs are obviously fertilized by these males, otherwise natural selection would have eliminated them from the population.

Chapter 10 deals with metabolic diversity, mainly the more obvious di-versity of metabolic processes that exist between say animals and plants, aerobic organisms and anaerobic organisms, etc. There are, however, examples of metabolic diversity within many species as well. Linked to genetic diversity is the occurrence of certain genetic diseases which al-ter some metabolic pathways. Though most of these are pathological to some extent, humans do survive in the face of some such conditions (often only with medical interventions and diet restrictions). Phenylketonuria is one such defect resulting from a defective enzyme. Without treatment this condition leads to mental retardation. Albinism also results from abnormal metabolic processes which fail to produce enough melanin, the pigment in the skin of "normal" humans.

Less abnormal or defective examples are also common. Some adults do not make enough of the enzyme lactase, used to digest lactose, a sugar found in milk products. These adults are said to be lactose intolerant and should not ingest significant amounts of cheese, milk, ice cream, etc. Other adults make more lactase and have no problem with lactose. Of course we all digest lactose just fine as infants because lactose is in our mother's milk, but many people lose this ability after childhood. In this case, it is the person who continues to produce lactase that is actually the more abnormal individual, at least when judged against the general pattern in mammals. We are the only mammal species in which some individuals continue to ingest milk and milk products past being weaned. Apparently in some populations, the continued production of lactase was "selected for," while many human groups retain the normal age-limited pattern of lactase production common to other mammals.

Our skeletal muscles have an amazing ability to switch from aerobic cellular respiration (see Chapter 10) to anaerobic lactic acid fermentation (a metabolic pathway common among some bacteria) when oxygen is limiting and the muscles are commanded to go on working. Most other tissues in our body do not have this ability. Also, some of our tissues (striated muscle and nerves) generate 36 ATP molecules from the respiration of a glucose molecule, while other tissues (liver, epithelial cells, etc.) generate 38 ATPs from a glucose molecule. This difference is due to a variation in the way an energy rich product of glycolysis (NADH) passes its energy (in the form of electrons) into the inner cavity of mitochondria. It's a complicated story which we need not go into here, but it serves as another good example of metabolic diversity within not only a species, but within the same organism.

Some typically photosynthetic species of the flagellate genus *Euglena* lose their chlorophyll under certain environmental conditions and become saprophytic heterotrophs, and in some instances this switching of metabolisms is irreversible, yet the heterotrophs would of course still belong to the same species as the typical photosynthetic individuals.

In Chapter 11 there is an overview of sensory diversity in humans and a few other selected groups. Humans have around 10 distinct senses, and perhaps some of these could be subdivided. Though a sponge or jellyfish would certainly have fewer, many organisms, including even unicellular ones, have a diversity of sensory abilities. Sensory abilities can also vary over a life cycle. Butterflies use vision and smell in finding a mate, while a caterpillar doesn't need to find a mate. Its senses are different and tuned to other needs. Of course there is also sensory variation with age. We know that in general children have far keener hearing and vision than people over 60.

In Chapter 12 behavioral diversity is discussed, and several examples of intraspecific behavioral diversity are included there such as behavioral differences between the sexes, caste-specific behaviors in social insects, age-related changes in behavior, distinct cultural differences between disjunct populations, etc.

Chapter 13 deals with diversity of life cycles across species, but life cycle diversity can exist intraspecifically as well. The nematode *Strongloides stercoralis* is a potential intestinal parasite of humans and a few other mammals. Quite distinct from most other parasitic nematodes, *Strongloides* can go through its life cycle as a free-living worm, or as a parasite. Free-living adults produce young which can grow into more free-living adults, *or* into young that develop into an infective stage larva which will not develop further unless they burrow into human skin to initiate the parasitic cycle. Likewise, parasitic adults produce young which, once outside the body, can develop into either free-living adults or infective-stage larvae. Cycling appears to be random and no mechanism has been discovered which determines which path of development the young will take. There are undoubtedly other examples of such intraspecific life cycle diversity.

These examples only hint at the vast amount of intraspecific biodiversity to be found in the millions of species of our planet. As mentioned previously, many other examples of intraspecific diversity will appear in most of the following chapters. Certainly most diversity of this type is yet to be elucidated, and much that we do know of, for whatever reason, is not traditionally thought of or discussed as being a component of biodiversity; but what else could it be?

5

Deep Biodiversity: Diversity "Above" the Species Level

... although *Naegleria* and *Entamoeba* seem to have much in common—
they are both parasitic protozoans that behave as amoebae for at least part
of their lives—they are more distant from each other, phylogenetically
speaking, than plants are from animals.

—Colin Tudge, *The Variety of Life*, 2000

This chapter concerns an important aspect of biodiversity, but one only
rarely mentioned as a basic component of biodiversity. If you were ab-
ducted by intelligent aliens, and back on their planet asked to give a
few examples that illustrated the other various life-forms of earth, which
examples would you give? Would you list and explain the differences
between a cardinal, a blue jay, a crow, a mockingbird, and a finch? It
is doubtful that even the most ardent ornithologist would offer up five
bird species as an illustration of the variety of life on earth. Surely any
thoughtful layperson would at least include a plant or two along with
some very different animals.

Almost any competent biologist could do better by including species
from the five recognized kingdoms such as: *Escherichia coli* (a bacteria),
amoeba proteus (a protist), bread mold (a fungus), red oak (a plant), and
rainbow trout (an animal). If allowed, the biologist would also try and
cover some of the major variations within the five kingdoms. If birds
were to be mentioned at all, only one would suffice to give an idea of
that type of animal. So, you could list five birds, all from the same class
of vertebrates, or you could list five organisms representative of the five
recognized kingdoms of life on our planet.

Though each of these responses contains five species, the second
would obviously be more representative of the variety of life on our
planet. I trust that this point is a "no brainer." Even the two most diverse

Table 5.1 The Seven Basic Levels of Classification with the Monarch Butterfly, Box Turtle, and Humans, Respectively, as Examples.

	Monarch Butterfly	Box Turtle	Human
Kingdom	Animalia	Animalia	Animalia
Phylum[a]	Arthropoda	Chordata	Chordata
Class	Insecta	Reptilia	Mammalia
Order	Lepidoptera	Chelonia	Primates
Family	Danaidae	Emydidae	Hominidae
Genus	Danaus	Terrapene	Homo
Species	plexipus	carolina	sapiens

[a] The phylum level is often termed the *division* level in the fungi and plant kingdoms.

bird species conceivable, perhaps a hummingbird and a penguin, are not nearly as diverse as a penguin and an oak tree. This admittedly extreme illustration clearly shows that species-level diversity is not the only type of taxonomic biodiversity worthy of consideration. Groups of species which are closely related and share numerous similar characteristics (such as birds) are not nearly as diverse as species from distantly related groups which have each evolved major suites of extremely different characteristics such as those distinguishing trees from tapeworms, mice from mushrooms, and bacteria from baleen whales.

Some biologists refer to this more fundamental diversity as "higher-level" diversity simply because when they list the levels of classification, they list the major groupings with the most divergent sets of characteristics above those with less divergent characteristics (Table 5.1). Because our classifications more and more reflect the phylogeny (evolutionary relationships) of groups of organisms, the so-called higher levels of classification represent older clades with older common ancestors (clades are phylogenetic groups that include all the species which speciated ultimately from a single extinct species—including that ancestral species). For example, the common ancestor of a phylum lived much farther back in time than the common ancestor of a class or order in that same phylum (Figure 5.1). For this reason, some biologists prefer to use the term "deeper" diversity, meaning that we must go deeper into the past (or more generations back) to encompass phylum-level differences than we need go to encompass family-level differences. Phylum-level differences will typically be far greater in terms of morphology and genes than family-level differences. Mockingbirds and earthworms are in different phyla, and their morphologies and gene makeup are very different. Mockingbirds and crows are in the same family, and their morphologies and gene-make are not nearly as divergent.

I share these concepts with my students when I cover marine invertebrates. At the species level, most invertebrate diversity is found in the terrestrial realm due to the fact that over two thirds of invertebrates are insects, and insects are almost exclusively terrestrial (with a few spending a portion or all of their lives in freshwater). But at any level equal to or higher/deeper than the class level, the vast majority of invertebrate biodiversity is marine, with 17 of the 35 animal phyla being exclusively marine, another five being mostly marine, and 12 phyla with at least some marine species. Only one animal phylum, the Onychophora (velvet worms), is entirely terrestrial. According to the fossil record even this phylum had its ancestry in the marine environment. Since the animal phyla are very old groups showing major differences (genetic, morphological, physiological, behavioral, etc.) from one to the other, the marine environment arguably holds the greatest amount of animal biodiversity, regardless of the greater number of terrestrial animal *species*, most of which are insects.

These greater and deeper differences between some species groups should be important to the way we view and manage the natu-

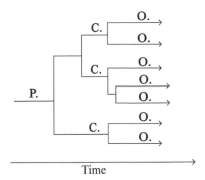

Figure 5.1 A simplified view of phylogenetic classification showing how older, more inclusive groupings include their newer and less inclusive groups in a "nested" manner, with orders (O.) nested into classes (C.), which themselves nest in phylum (P.) groupings. Most actual phylogenies would include far more branching than that show here. The animal phylum Mollusca for example contains seven classes, and one of those classes, the Gastropoda, is further divided into seven orders. (*Figure by author*)

ral world. The discovery of a new species is not really big news. Several new species are discovered and described every week. These are usually new species of insects, worms, protists, etc. with the occasional fish or mammal popping up. The discovery of a new class or phylum, on the other hand, is a very big deal.

In 1995, a new phylum of animals, the Cycliophora, was described from the marine environment. These small and very different animals live as symbionts (apparently commensals) on the mouthparts of lobsters (talk about microhabitats!). This discovery made the national news media, and almost everyone in academic zoology was soon aware of this significant find. Likewise, losing a rare species of finch would not be as great a loss as losing the common Atlantic horseshoe crab (*Limulus polyphemus*). There are over 120 species of finches across the globe, as well as many other closely related bird species in the same order, but there are only four species of horseshoe crabs worldwide making up a class-level taxon (Merostomata) in the great phylum Arthropoda. The genes, morphology, physiology, and ecology of horseshoe crabs are unique, ancient (over 360,000,000 years old), and rare features in the world of living things. The same cannot be said for any one species of finch.

Of course we would rather not see any species meet extinction through our influence, but those with sparse and ancient ancestries could be viewed as more precious than those of more recent and highly speciated (speciose) groups. By the same token, ecosystems which include more of this deep diversity are more unique and irreplaceable than those made up of species which are found elsewhere, or are members of speciose and widespread taxonomic groups. Measures of this deep diversity should arguably be included and considered when making a case for the preservation or conservation of particular natural areas and ecosystems.

It might be obvious that the deepest levels of diversity would be found in the oldest group of organisms. Recently this has been found to be true of the prokaryotic organisms which we have placed in the kingdom Monera. Prokaryotes (organisms whose cells lack nuclei) were the sole inhabitants of earth for around 2,000,000,000 years prior to the appearance of nucleated cells. Due to careful comparisons of DNA and RNA molecules, it is now believed that the deepest major division in biodiversity is between two huge prokaryotic groups, the bacteria and the archea. These two groups are referred to as domains. All the eukaryotes make up a third domain, the eucarya (Figure 5.2). These three groups are the highest or deepest clades that biologists have been able to detect.

This causes a slight problem if we formalize the domain level as a taxon above the kingdom level. The problem is that we would have two domains in the single kingdom Monera. This is like having two phyla in the same class, which is an impossibility. It has already been suggested by several workers that we should formally adopt the three Domains and then proceed to establish several prokaryote Kingdoms within the Bacteria and the Archea. This would be a major change in the way we have separated and named our higher taxonomic ranks. It will certainly take some time before the outcome of these ideas and proposals will be settled. At any rate, our knowledge concerning this deep biodiversity is

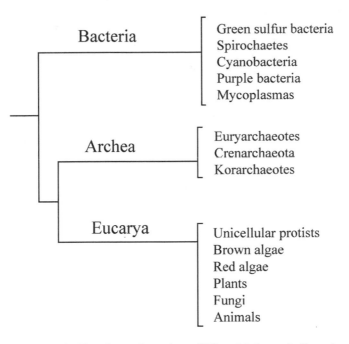

Figure 5.2 The three domains of life, which are believed to be the oldest (deepest) major divisions in the phylogeny of life on Earth. Not all known groups within each domain are listed. (*By author after various sources*)

now growing at a rapid rate due to new fossil finds, new ultrastructure studies, new developmental studies, and new species discoveries (which sometimes hold missing pieces of a "phylogenetic puzzle").

The most powerful tool for clearly "seeing" this deep biodiversity comes from the new techniques of comparing the biochemistry of species at the level of proteins and nucleic acids (DNA and RNA). Evolution after all is defined as a change in gene and allele frequencies within lineages, and these changes don't all equate to obvious outward changes in morphology. Comparing DNA, RNA, and proteins directly is the most unbiased and powerful tool for establishing just how deep the distinctions between Families, Classes, Phyla, etc. really go. Research into this deep biodiversity is currently a "hot" area of research and theory which is unlikely to cool in the near future. Biologists really do want to know how groups are related in terms of who is who's closest relative, how deep is the division between the higher taxa, and how best to name groups in a way that reflects their phylogenetic relationships—more on this topic in Chapter 15.

6

Genetic Diversity

In this alphabet (of nucleotides and amino acids) can therefore be written all the diversity of structures and performances the biosphere contains.
—Jacques Monod, *Chance and Necessity*, 1971

Young or old, all living species are direct descendants of the organisms that lived 3.8 billion years ago. They are living genetic libraries.
—E. O. Wilson, *The Diversity of Life*, 1992

Genetic diversity is typically the second listed/mentioned component of biodiversity. Though evolution (itself working mainly on genetic differences) is the ultimate factor behind species diversity, the main proximate factor behind the distinctiveness and diversity of species is the current genetic disparity between those species. The degree of distinctiveness between species in terms of morphology, behavior, metabolism, etc. is broadly correlated with the degree of their genetic disparity. Any two species of finches native to the Galapagos Islands are only slightly distinct from one another because they are recently descended from a common ancestor and are thus identical in a very high percent of their DNA or genes.

Either of the two species of finch in question would be drastically different in most of its traits from yeast, because finches and yeast share a much smaller percent of identical DNA, though they do share some. Yeast and finches do share a common ancestor, but one which existed over 1 billion years ago, and their respective genetic allotments have diverged greatly over that long stretch of intervening time. It is relatively recent knowledge that animals and fungi shared a common ancestor later than either did with plants. This knowledge resulted from the recent extensive genetic analysis and comparison among species within each group.

The general rule that more obvious species differences correlate with genetic disparity has its exceptions. In sibling species, genetic disparity seems to have outpaced the evolution of morphological and other "outward" distinctiveness. This is often true in some groups of microorganisms (especially unicells) that have less morphology available for divergence. Also, some species that appear to us to be quite disjunct can in fact be quite similar in their genetic makeup. Humans and chimps differ from one another in only around 1.5 percent of their genetic makeup, yet their respective morphology and behavior seem to us to vary by a much greater degree (perhaps due to our biased impressions of ourselves). Some parasitic copepods differ so drastically in their morphology from free-living copepods that they might believably belong in different phyla or kingdoms (see Figure 15.1), even though they are in the same Class and share a majority of their genetic makeup in common.

One genetic parameter that can and does vary between species is simply the number of genes present in the cells of individuals of those species. Human cells are now estimated (posthuman genome project) to have around 25,000 different genes (with two pairs of each type in a diploid cell—one from each of the two parents). The well-studied nematode worm *Caenorhabditis elegans* has roughly 17,000 genes. The prokaryotic Bacteria and Archaea typically get by on less than 5,000 genes, and viruses (which don't actually qualify as living organisms) usually contain fewer than 100. It may come as a surprise that our 25,000 genes do not define the upper limit for gene content. A few seemingly simpler organisms such as certain species of salamanders and flowering plants can contain well over 200,000 genes (though many of these may be redundant or noncoding). As a general rule, simpler organisms contain fewer genes than more complex ones, even though there seem to be numerous exceptions. Remember too that to date we only have gene counts for a small percent of the earth's species.

It is certainly possible that a certain species of plant and a certain species of animal might turn out to have exactly the same number of genes. Obviously then, it is not simply the number of genes that determines the characteristics of a species but the types of genes that are present. This is clearly illustrated by polyploidy in some salamanders (and many other groups). Two species of salamanders can appear very similar in morphology, behavior, and chemical makeup, yet one of the species can have twice as many genes as the other because its ancestor underwent polyploidy such that it now has essentially duplicate copies of all the genes in the nonpolyploid relative. It is essentially 4N rather than 2N, but once a species has completed such a transformation, we still say it is 2N, but simply with twice as many chromosomes as other related species.

There would likely be some genes that are identical or near identical between a plant and an animal, but most would be quite different. Some plant genes would code for the enzymes involved in the unique metabolic steps (reactions) of photosynthesis. These genes would most likely not be present in an animal. Some animal genes would code for assorted neurotransmitter molecules, which plants would not possess since they have no nervous systems. The more closely related two species are, the more genes they would share due to their common ancestry. The more distantly related, the less similar their total gene similarity would be.

Genetic diversity is the underlying "program" behind much of total biodiversity. Genes determine to a large extent the types of molecules produced by cells, the types of cells found in an organism, the types of metabolic activities the cells engage in, the general morphology of the organism, and the lifestyle and potential behavioral repertoire of the organism. Genes can vary between members of a species as Mendel discovered, and as Darwin recognized phenotypically (the outward expression or result of gene coding). It is of course this intraspecific genetic variation between individuals that allows natural selection to forge adaptations, thus perfecting the "fit" of the species to its environment (and giving the potential to evolve).

Genes can also vary within individuals. This is because many genes exist in two or more variant forms within a species. These variants of a gene are called *alleles*, and diploid organisms like ourselves can carry two alleles of the same gene (one inherited from each parent), a condition referred to as *heterozygous*. A person who has type AB blood is heterozygous because she carries two forms of the same gene and two forms of the gene product—an A form and a B form. A person who has type O blood has two genes which are identical (the *homozygous* condition); both are O form which is the third known allele of that same gene.

Some genes have many alleles present within their respective species. Since every human cannot be sampled to check for variant alleles at each of our 25,000 genes, we really don't know what percent of human genes have allelic variation in the entire species, especially considering that some alleles could be extremely rare. If we as a species had alleles for a third of our genes (which some earlier estimates suggested), then 8,000 of our 25,000 genes would have at least two variant forms, and of course some have three as in the A-B-O blood group, and some certainly have more. Since the percent of our genes which have alleles somewhere in the human species is most likely higher than 33 percent, and since some of these genes have several alleles, there are almost certainly far more than 25,000 kinds of alleles in the human species. If similar trends are present in other species, the number of variant genes in all species of living organisms (most of which we haven't even discovered yet!) is such an astronomical number that even to attempt an estimate would be unhelpful, since no one could comprehend it. Genetic diversity clearly exceeds biodiversity at any other level, except that of individuals (Chapter 4).

Alleles are usually discovered by their slightly different gene products; proteins or polypeptides that vary from one another by one or more amino acids (the subunits of proteins). Investigations of gene products over the past 50 years have revealed a great amount of genetic and molecular diversity which had been previously unknown, because many variant proteins do not result in obvious macrolevel phenotypic character variations. Through the technique of gel electrophoresis, many variant forms of enzymes have been discovered in a wide array of species. Isozymes are variant enzymes that are the products of variant alleles of the same gene. Allozymes are variant enzymes produced by variant copies of duplicated genes (listed as one of the sources of diversity in Chapter 2). Many of these variant molecules had remained undetected prior to the advent of gel electrophoresis and other new techniques for deciphering genetic and molecular diversity.

Today through gene sequencing we can find alleles that vary in some of their nucleotides, even where the gene product is the same as that of an alternate allele. Since the triplets AAT and AAC both code for placement of the amino acid Leucine into a peptide chain, the gene products would be identical while the two alleles would in fact vary (and thus be alleles). There are many such triplet synonyms in the generic code. Triplets are the three-base "words" in genes/DNA that code for amino acids in proteins. This is the least obvious type of neutral genetic change or evolution, where the DNA changes or diversifies through mutation, but its coded protein products remain unaffected. Such slight variations might be viewed as unimportant and not worthy of mention, but these nucleotide changes and other less-obvious molecular variations do provide information of great value in the work of elucidating phylogenies (see Chapter 15).

Of course DNA changes often do lead to amino acid changes in proteins, and some of these can also be neutral in effect. In the case of variant forms of the same enzyme, an amino acid change would likely only have a significant effect if the change altered the so-called "active site" of the enzyme. This is the business end of the enzyme where it bonds briefly to a substrate molecule to affect some change in that substrate. If the amino acid change occurred on the "backside" of the enzyme away from the active site, the result could easily be an equally effective enzyme with a slight difference in one of its building blocks, in short—a neutral variant.

If the triplet AAT mutated to AAA, then Phenylalanine, not Leucine, would be the amino acid coded for insertion into the protein. Depending on the protein and the position in the protein of this change, the result could either be a neutral effect, a beneficial effect (the protein "works" better), or a harmful effect (the protein becomes less effective or totally nonfunctional). The most likely result of a random mutation would be the third possibility—a harmful effect. Randomly altering a smooth-running machine most often results in machine failure. Probably the second most likely outcome would be a neutral effect, with really beneficial mutations occurring only rarely.

The harmful mutation would typically be short-lived in the population, or at least remain at a low frequency, due to natural selection. The neutral mutation would more likely be retained in the population to contribute to biodiversity at the genetic level, though its frequency too would likely remain low in the population, especially in a large population. Neutral mutations affecting amino acid makeup have most definitely occurred in countless proteins in most, if not all, species. This is why we can trace evolutionary changes in cytochrome c, a mitochondrial protein important in cellular respiration. Cytochrome c contains just over 100 amino acids (104 in vertebrates). The many neutral changes that have been detected that affect amino acid makeup do not apparently affect the basic function of the molecule, which is to serve as part of the electron-transport chain in mitochondria, but these changes have certainly created diversity. Table 6.1 shows the degree of difference in the cytochrome c of various species. This is another case where even neutral diversity is useful, not to the organisms possessing those variant forms of cytochrome c, but to scientists who can read the record of these changes to aid in deciphering evolutionary relationships, an important goal within the biological sciences. Once again, it is the differences in genes that result in this "higher" level of variation within larger taxonomic groups.

A few genes code for ribosomal RNA molecules (rRNA) used in the construction of ribosomes. Ribosomes are the "workers" which in a sense read messenger RNA molecules and construct proteins accordingly. Also, a few genes code for transfer RNA molecules (tRNA) which

Table 6.1 Human Cytochrome c is Shown in Comparison to That of Other Species

Humans and chimpanzee	0
Humans and rhesus monkey	1
Humans and horse	11
Humans and turtle	15
Humans and frog	18
Humans and fruit fly	24
Humans and mold (Neurospora)	40

Note: Cytochrome c is a protein typically composed of 104 amino acid units. The numbers in the second column represent the number of amino acids in Cytochrome c that now differ between humans and the organism listed. (Assembled from various sources.)

"grab" and transfer amino acid molecules to the ribosomes for protein construction. Most of the coding genes code for proteins by way of messenger RNA molecules (mRNA). This mRNA is essentially the blueprint for constructing a protein, the blueprint "read" by the ribosome. A review of this process can be found in any high school or freshman-level biology text, but the main point for now is that gene diversity translates into molecular diversity (rRNAs, tRNAs, proteins, etc.), and yes, genes are molecules too.

Another example of molecular diversity in humans involves what are known as the facilitated diffusion glucose transporters; molecules which function in the outer cell membranes of cells to transport glucose (a required nutrient) into cells. There are four types of these glucose transporters which are all undoubtedly variants of some ancestral glucose transporter whose coding gene has been duplicated and subsequently diversified by mutation and natural selection. These four related glucose transporters now serve different needs in the bodies of complex organisms such as ourselves.

The four glucose-transporter variants are known as Glut 1, Glut 2, Glut 3, and Glut 4. Glut 4 is only found in muscle and fat cells and is the only transporter responsive to insulin. As insulin levels rise, more Glut 4 is added to the muscle and fat membranes, and more glucose is transported into these cells thus lowering blood glucose levels. Glut 3 is found mainly on nerve cells and is not responsive to insulin. Glut 2 is found on liver cells and intestinal epithelial cells. Most cells of the body also contain Glut 1. The four transporters vary also in their respective efficiencies of glucose transport or intake into the cell.

Though all the cells in a multicellular body contain the same complete set of genes, different cells diverge during development into distinct tissues. Subtle influences of embryonic cell movement, placement, and extracellular environment determine which cells develop into which tissues through selective activation of a subset of the cell's total genes. In a nerve cell, genes that would code for muscle cell or fat cell structure are simply ignored, and vice versa.

So, from molecules, to cells and tissues, to individuals and species, diversity is most directly the result of genetic diversity. This is not to ignore that some diversity can result from environmental factors (such as the tissue divergence in embryos mentioned earlier). We certainly recognize how diet, exercise, sun exposure, education, culture, etc. can affect human diversity in terms of both morphology and behavior, and there are many other examples from other organisms, some of which will be covered in Chapter 9 which deals with morphological diversity.

Unlike the less obvious differences in cytochrome c, blood groups, and other cryptic biomolecules, there are countless other genetic/molecular differences which collectively result in morphologies as distinct from one another as those of mushrooms and mosses, amoeba and ants, and leeches and lilies. Whether it is small differences between two species of cricket, or large differences between a fish and a daisy, they are largely the result of genetic differences. Genes also result in variation above the species level in particular homologous molecules like cytochrome c (human and chimpanzee cytochrome c are identical), and at all other levels of variation between the higher taxa.

Because each species has at least a few genes unique to that species, the loss of any species to extinction results in the loss of unique genes and gene products. As was mentioned back in Chapter 5, losing a species that has several closely related species in the same genus is less costly in terms of gene loss as compared to losing a truly unique species with no close relatives. Within the animal kingdom there is the phylum Placozoa that contains only a single very unique and primitive marine species (*Trichoplax adhaerens*). Being the only species in its phylum, *T. adhaerens* has no close (not even remotely close) relatives. This taxonomic remoteness means that the genome of *T. adhaerens* is exceptionally unique, such that the loss of this one species would be a very significant loss in terms of thousands of genes that exist nowhere else in nature. Whether or not these genes will ever have any pragmatic usefulness to humans through use in biotechnology, they and their gene products should be valued because of their uniqueness, their antiquity, and their information content concerning animal evolution.

We often think of DNA and genes as essentially the same thing, but this is not the case. Genes must typically code for something (a protein product, rRNA, tRNA, etc.) in order to be considered genes, but much of our DNA does not code for any molecular product at all. Some stretches of DNA are thought to be regulatory in terms of separating genes, serving as attachment sites for polymerases, serving as on/off switches, etc. Some of our DNA consists of vestigial "genes" which no longer code for molecular products. When a trait is lost over evolutionary time, what is often actually lost is the ability to turn a gene on and so produce a gene product. The gene itself is most likely still present but inactive. Such "silent" genes undoubtedly get passed on generation after generation at very insignificant cost to the species. These genes are like the old games, clothes, tools, etc. stored in many people's attics or basements. They haven't been discarded, but they aren't being used anymore. Vestigial genes likely make up much of the so-called "junk DNA" in our chromosomes.

Then there are the so-called "pseudogenes." Pseudogenes are duplicated copies of functional genes (duplication of genes was mentioned in Chapter 2) which have built up mutations to the point that they are now silent and nonfunctional. No one knows how much of our DNA exists as pseudogenes, but it is probably a significant percentage. There are also stretches of DNA which are not genes, not regulatory, and not vestigial. At least some of this extra DNA is "parasitic DNA." As Richard Dawkins has written in *The Selfish Gene* and *The Extended Phenotype*, the goal of DNA is simply to get itself passed on in time to future generations. Many examples have been found of units of DNA which are quite good at getting themselves passed on, without giving any benefit whatsoever to the bearer (the individual or species). *Transposons* are a prime example of such parasitic DNA. Though it is a complex story beyond the scope of this book, transposons may even be the source of some viruses, which typically have harmful effects on the species that spawned them.

Are vestigial genes, pseudogenes, and parasitic DNA a part of biodiversity? Yes, of course they are. Their relative importance may be low in most cases, but at least they teach us important lessons on how evolution and "nature" work. They, like our functional genes, can also be deciphered and compared across species to aid in our understanding of the phylogenetic relationships of those species. Closely related species will have inherited some of the same vestigial and parasitic DNA from their common ancestor.

Genetic diversity within a species is a desirable situation because genetic diversity is essential for species adaptation to changing environments (evolution). A lack of genetic diversity could allow a new disease, a severe drought, a long-term climatic change, a new competitor, a changing food base, etc. to endanger a species and even drive it to extinction. Biologists worry when a species is reduced to low numbers because the level of genetic diversity is likewise reduced in that species, thus making it more vulnerable to extinction. We know that our major farm crops lack adequate genetic diversity, which makes them susceptible to a variety of environmental insults that would pose no problem for a more genetically diverse species. It thus requires a lot of fertilizer, irrigation, insecticides, herbicides, etc. to protect our crops and insure their survival. They are really poor survivors in the absence of our care and protection. This explains the importance of preserving some of the genetically more variant wild populations of the ancestors and close relatives of our domesticated crops and animal species. They can serve as reservoirs of genetic variation, variation which can be reintroduced as needed into our genetically-impoverished, but vitally important, crop varieties to "strengthen" and renew them from time to time.

When we discover that some chemical from a wild species has potential cancer-fighting properties, it is the gene or genes that code for that chemical that are the focus of attention, not the species itself. More and more we are able to isolate that gene and implant it into a faster-growing and faster-producing species to harvest that chemical, but remember that we would not have discovered the gene without the source species. We must keep this in mind as more and more species face extinction. As more species die out, we are undoubtedly losing untold treasures in the form of unique chemicals coded for by genes unique to those species. Some of those genes/chemicals will undoubtedly have great potential value for humanity, assuming they are not lost. Genetic diversity is definitely valuable and worthy of preservation, and the best way to preserve it is to make sure that species don't go extinct because of the short-sighted actions of our own species.

7

Diversity of Ecosystems and Ecological Interactions

Biodiversity occurs not only at the level of species, but embraces both species subunits and ecosystems.

—Norman Myers, The Rich Diversity of Biodiversity Issues, in *Biodiversity II*, 1997

Vast reservoirs of diversity also exist at other levels of biological organization, from the letters of the genetic code within local populations to ecosystems comprising large numbers of species.

—E. O. Wilson, Biodiversity: Challenge, Science, Opportunity, *American Zoologist* 34 (1994): 5–11

The obvious fact that some ecosystems are quite distinct and different from others is the reason why ecosystem diversity is recognized as one of the "big three" categories of biodiversity. The definitions of the terms *community* and *ecosystem* vary and are conceived somewhat differently by different ecologists. The term community typically refers more narrowly to only the living components and their interactions within a somewhat defined area such as a lake, a hardwood forest, a coral reef, etc. The term ecosystem more often includes both the biotic *and* abiotic components and all their various interactions. The abiotic or nonliving components include climatic factors, altitude, salinity, pressure, dissolved oxygen level, substrate type, etc. Although no ecosystems are entirely self-contained and separate from other ecosystems, it is sensible and useful to speak of a coral reef ecosystem (or community), a desert ecosystem, a lake ecosystem, etc.

Ecosystems can vary in their species makeup (both the actual species present and the total number of species), in the types and complexity of biotic interactions, and in the nature of their abiotic interactions. Though there will likely never be complete agreement on a classification or listing of ecosystems on our planet, it can certainly be argued that there are a great many distinct ecosystems now extant on our planet. There are the commonly mentioned "biomes" which are covered in most general and ecology textbooks. Terrestrial biomes include desert, tundra, taiga, temperate forest, grassland, tropical rain forest, tropical savanna, chaparral, and mountains. Some sources subdivide these even further. It must be recognized that a tropical rain forest in Costa Rica is a completely different rain forest from those in the Congo basin in Africa, and these are different from those in many other areas of the world. Each has its own unique species makeup and interactions. Consider islands. Most of the world's islands are individually unique in their list of species and particular ecological interactions. Even in the closely grouped Galapagos Islands, each island has its own endemic species and subspecies with their own unique set of ecological interactions.

Freshwater ecosystems include lakes, ponds, rivers, streams, swamps, and springs. There are brackish water estuaries (with many variations) and the huge world of marine ecosystems. There are probably as many marine ecosystems (somewhat equivalent to the terrestrial biomes) as there are terrestrial ecosystems. Just because "the marine environment" is all covered with saltwater no more makes its resident ecosystems similar than terrestrial biomes are made similar by being covered by air. Coral reefs, intertidal communities, hydrothermal vent communities, abyssal plain communities, sand flat communities, kelp forest communities, mangrove communities, sea grass communities, meiofaunal communities, open water or pelagic communities, etc. are all distinctly unique ecosystems with basic component and functional differences.

A great many more ecosystems never get mentioned in most sources because they are somewhat obscure or simply escape the attention of most writers. It could be argued that there are at least thousands of distinct and different ecosystems on our planet, even though a more general term might be inclusive of several somewhat similar ecosystems. For example we can speak of a desert ecosystem (or biome), but there are many distinctly different desert ecosystems in the world, some with truly unique sets of resident species and abiotic parameters.

Likewise, lakes are not all identical in their species makeup, their biotic and abiotic interactions, their seasonal changes (or lack thereof), etc. An alpine lake is a completely different type of ecosystem from the Great Salt Lake in Utah. It could be argued that most lakes on this planet constitute distinct ecosystems with their own unique combination of biotic and abiotic interactions, even though certain generalizations will apply to many or all of those lakes.

Another example would be the ecosystems that exist in the digestive tracts of large animals such as buffalo, deer, bears, etc. Typically hundreds of species of microorganisms and several symbiotic animals live in the digestive tracts of any one large animal species, and some of those will not be found in any other host species. Since the diets of such animal species vary, the "flora" (bacteria) and fauna of the gut of each animal species will be distinct in several ways. Many might question the inclusion of gut communities as distinct and worthy of mention, but why not? Though they may be patchy, with each patch (host animal) being eliminated eventually, they certainly do seem to fit generally the standard definition of an ecosystem.

Likewise it has been found that most tropical trees contain a surprising diversity of insects (and undoubtedly many other organisms) that seem to be unique to that type of tree. Could it not then be argued that each of these tree species houses a unique ecosystem of interacting species, in the same way that each lake contains a unique ecosystem? In essence what we find are "ecosystems within ecosystems," because those large animals and tropical trees are each in turn members of a larger ecosystem.

In the oceans, large sponges often serve as habitat for a variety of other organisms. I have seen literally hundreds of small polychaete and other worms emerge from a marine sponge the size of a tennis ball (after sitting for a time in a warm lab in still water). Small crustaceans, brittle stars, and sea spiders also regularly inhabit certain sponges along with a great variety of endemic symbiotic microbes. If these sponge inhabitants vary from one sponge species to another, interact in predictable ways, and live their lives almost entirely in association with that sponge species, then isn't there a unique ecosystem in almost every species of sizable marine sponge?

Let's move on to another point, which is that several entirely new ecosystems have been discovered in the last few decades. Some of these "new" ecosystems are so different and unique compared to those already known that they have vastly expanded our understanding of ecosystem diversity. Consider the several discoveries of chemosynthetic ecosystems, the first of which was discovered in 1977 in deep water off the west coast of Equador. This deep-sea hydrothermal vent area was not expected to contain significant life, but when the lights of the submersible *Alvin* were cast over the immediate vent area, life was everywhere. Most of the worms, clams, jellyfish, and fish inhabiting this area were new to science. Most significantly, it was discovered that this whole community of animal life was not fueled by photosynthesis, but by the process of bacterial chemosynthesis (see Chapter 10). Several species of bacteria (also mostly new to science) were chemically manipulating inorganic substances (carried in the hot waters which spewed from the hydrothermal vents) in such a way as to release energy. This chemical energy replaces sunlight for the bacteria. They are able to harness this chemical energy to build organic molecules needed for growth and maintenance (chemosynthesis). The animals in this ecosystem all live directly or indirectly on the productivity of the resident chemosynthetic bacteria.

The finding of this and several subsequent chemosynthetic ecosystems, one of which was found to exist in a cave in Romania, was totally unexpected, and expanded in significant ways our knowledge of ecosystem variety. The many marine chemosynthetic communities vary dramatically in their species makeup and in the chemosynthetic processes that support each community. Some of these communities are supported by methane, some by natural oil seeps, some by hydrogen sulfide, and some by elemental hydrogen and carbon dioxide. So although they can be given a common term like marine chemosynthetic communities, some are as different from one another as most of the traditional terrestrial biomes, which at least share the same energy source—the sun.

Another ecosystem that has only been elucidated within the last few decades is the meiofaunal ecosystem. Meiofaunal ecosystems are those which exist *in* marine bottom mud and sand. A whole host of animals, protists, and prokaryotes are now known to have specialized for life between the grains of bottom mud and sand. They are obviously very small, often slender organisms that can occur by the millions per square meter of bottom sediment. A couple of newly discovered animal phyla, the Loricifera and Gnathostomulida, were found to be regular members of this previously overlooked, though extremely common habitat.

In some islands of southern Japan several sizable underwater caves, which vary from near to total darkness, have been discovered to house communities of previously unknown species of bivalves, gastropods, sponges, corals, bryozoans, protists, etc. There are certainly a great many unique cave ecosystems both below and above water that await discovery and study.

Another relatively newly recognized set of ecosystems or habitats are the thousands of extensive cliff faces in various places around the world. Depending on the grade or steepness, mineral content of the underlying bedrock, sun and wind exposure, depth and moisture-holding properties of the soil, etc. cliff communities can vary greatly from nearby communities above or below the cliff face. Many plants are adapted for such extreme habitats and are found nowhere else. Many of these are "dwarf" species and subspecies, and unique plants of course typically have unique animals species (especially insects) associated with them. In terms of species diversity, cliffs are beginning to be recognized as disjunct reservoirs of relatively high biodiversity. A cliff that faces north can have a distinctly different community from an opposing south-facing cliff, even though the two may be separated by only a few hundred feet. Their distinctiveness can be enhanced if they also vary in other parameters such as grade.

Bacteria formed the first communities or ecosystems on this planet. These communities functioned for close to 2,000,000,000 years before the appearance of eukaryotes. Bacteria are the only group of organisms capable of forming (by themselves) a complete ecosystem due to their varied metabolisms and their exclusive roles in the biocycling of most elements and compounds. Today most ecosytems include both prokaryotes and eukaryotes, but there are still a few ecosystems where prokaryotes form the major or only biotic components.

Though there are often a few small arthropods and other invertebrates inhabiting large and deep caves, we now know that the main inhabitants of many caves are communities of chemosynthetic bacteria that form coatings or hanging masses on the cave surfaces. They derive their energy in a manner similar to those bacteria of the deep-sea hydrothermal vent communities mentioned earlier. Across the bottom of some Antarctic lakes and in some hot spring pools live mats or "biofilms" of several species (communities) of extreme-adapted bacteria and archea which are the only living inhabitants of these isolated and inhospitable waters. Stromatolites, which still exist in some marine habitats, are really insular communities of several species of bacteria that interact very little biologically with the larger surrounding ecosystem.

Over the past decade it has also become clear that there exist huge microbial ecosystems of chemosynthetic prokaryotes living deep within the earth's crust at depths of up to 5 kilometers or more. Some of those involved in these studies suggest that there may be as much collective biomass in these subterranean microbes as in the more accessible and obvious above-ground ecosystems. Though these microbes live extremely slow lives, probably with few biotic interactions, they must certainly constitute a simple and extensive ecosystem which differs dramatically from the more familiar ones above the surface. It has even been suggested that the first life on earth may have been similar to these modern subterranean microbes.

As mentioned in Chapter 2, ecosystems are one of the "higher" recognized levels of organization and structure in the living realm. Because ecosystems vary so dramatically in their components, interactions, and complexity, we can confidently and intelligibly speak of the biodiversity of ecosystems. The more you know and think about nature, the more ecosystems you are likely to recognize; some are large, some are small, and many nest within each other as in some of the earlier examples.

DIVERSITY OF ECOLOGICAL INTERACTIONS

What of the specific individual ecological interactions? How many categories and variations occur among these interactions? Obviously there is much biodiversity residing here as well. We know some of the larger categories from basic ecology; categories such as commensal, parasite, predator, scavenger, pollinator, herbivore, etc., but under any one category there can be much diversity. For example, here are a few of the many possible ways of being a predator:

- Chasing down prey: cheetahs, wild dogs, porpoise, squid
- Luring prey in: angler fish, alligator snapping turtle
- Digging up/out prey: the aye-aye, woodpeckers, anteaters
- Trapping prey: Orb-web spiders, Venus fly traps
- Sit-and-wait predation: trap-door spiders, mantids, corals
- Shocking prey: Electric eels
- Disabling prey with ejected "glue": velvet worms
- Wandering opportunistic predation: Tarantulas, most jellyfish, some amoeba
- Tool-use in prey capture: "Fishing" for termites in chimps, "spitting" water at prey on vegetation overhanging the water in the archer fish
- Cleaning other animals of parasites: cleaner shrimp, cleaner wrass, ox-peckers
- Filtering prey from water: baleen whales, basking sharks
- Eating your mate: some spiders and mantids (part-time)
- Directed group raids: some ant colonies against other ant nests
- Poisoning prey: Blue-ringed octopus, pit vipers

There could be overlap in the examples above, some of which are general strategies, while some are more specific prey-capture techniques. If one notes the general predatory strategy along with the specific capture/killing method and the exact feeding method, one could produce a very long list of distinct predatory interactions found in nature.

It would be far beyond the scope of this book (or any book) to detail all the various ecological interactions that occur in ecosystems (almost certainly an impossible task). Nevertheless, it might be instructive to throw out a few illustrating examples in the hopes that they will remind us of the great diversity that exists among such interactions.

- Pollination of flowering plants by honeybees, in return for nectar and pollen.
- Aphids which suck sap from plants.
- Pearl fish which spend much of their time within the rectums of sea cucumbers, presumably for protection.
- Certain mosquito and other fly larvae which live until adulthood in the liquid within pitcher plants.
- Leeches which suck blood from various vertebrate hosts.
- Cattle egrets which follow alongside large grazing ungulates waiting for insects to be flushed into the air where the egrets can capture them.
- Mutualist bacteria and protists which aid in cellulose digestion in the digestive tracts of ungulates, termites, and shipworms.
- Epiphytes such as some orchids and bromeliads which grow attached to other larger plants.
- Insects which insert their eggs into certain plants which are then stimulated to form galls around the egg/larvae.
- Competition between plant species for sunlight, water, soil nutrients.
- Mutualistic nitrogen-fixing bacteria living in the root nodules of legumes.
- Chemosynthetic bacteria living as mutualists in the bodies of the giant beardworms located at many of the marine hydrothermal vent areas.
- Animals such as chimps which ingest certain plants not for food, but for medicinal purposes (much evidence now that this occurs).
- Plants which produce seeds with coats or fruits which stick to animals for dispersal, or are eaten and pass through the animal unharmed and dispersed.
- A vervet monkey escaping a leopard by climbing a tree.
- A skunk defensively spraying a potential predator.
- Young chimps playing with young baboons (interspecific play).
- Hermit crabs attaching sea anemones to its gastropod shell home for camouflage and defense.
- Marine pelagic amphipods of the genus *Hyperiella* holding and carrying on their backs sea butterflies (*Clione sp.*) to benefit from their antipredator chemical defenses (the sea butterflies taste bad).

- Certain nudibranchs feeding on Cnidarians and then using their stinging nematocysts second-hand for their own defense, contained in their dorsal finger-like cerata.
- Interactions of parasites and hosts where the parasite somehow alters the behavior of the host in a way which increases the parasite's fitness while lowering that of the host, as in the fluke/snail example covered in Chapter 13.

This is but a small sampling of nature's interspecific interactions (interactions between different species), but intraspecific interactions (interactions within the same species such as competition, courtship, etc) are also ecological interactions, as are abiotic interactions with environmental factors such as temperature, wind, salinity, gravity, mineral composition of soil, light, pH, barometric pressure, moisture, etc. There are also many other biological interactions which aren't typically thought of as ecological such as an allergy to plant pollen or a strong reaction to poison ivy. Ecological interactions are more diverse than species because each species engages in several different types of ecological interaction. Though some of these can be grouped into general categories like pollination, the details of who, how, when, etc. are specific to the particular plant and pollinator species, thus the incredible diversity—the biodiversity of ecological interactions.

8

Four Billion Years of Temporal Diversity

The number of specimens in all our museums is absolutely nothing compared with the countless generations of countless species which certainly have existed.
—Charles Darwin, *The Origin of Species*, 1859

Organisms diversify into literally millions of species, then the vast majority of those species perish and other millions take their place for an eon until they, too, are replaced.
—George Gaylord Simpson, *This View of Life*, 1964

. . . there are several thousand million years of evolutionary drama behind everything that moves and breaths.
—Colin Tudge, *The Variety of Life*, 2000

We walk the same planet where millions of years ago all manner of strange and diverse creatures and organisms lived out their lives. Many of these unique and diverse organisms were as different from our modern organisms as some modern organisms are from one another, and there are still plenty of strange organisms around today. Though many of the earth's past inhabitants would be at least somewhat recognizable to an undergraduate biology student, many were strange and alien compared to anything we could find in today's world.

You can easily find the statement in many biology texts that over 99 percent of all the species that have ever lived on this planet are now extinct, having left no direct descendants. This is a statement most biologists understand and accept as most certainly true, even though we do not have an objective record of anywhere near such a high number of extinct species. The number of known extant species is around 2,000,000; so according to the statement earlier, at least 200,000,000 species have lived and are now extinct. Remember too that the number of *known* extant species is but a fraction of those which actually exist, and it is more likely that billions of species have lived and are now extinct.

How can such a claim be viewed as true? One reason why this claim is likely true is the branching pattern of each evolutionary line. Speciation events are very common over the long stretch of evolutionary time, and most of the resulting species and clades died out without contributing to an ongoing lineage. Also, some of those extinct clades were quite sizable. We have documented around 900 species of dinosaurs. Although one small branch of the dinosaur lineage continued on to become the bird clade (and we probably haven't discovered the exact origin of that branch), all the other dinosaur lines and species are now extinct. Though we still have a few large monitor lizards around, in size and probably also in behavior nothing today even comes close to the undoubtedly amazing sight a living *T. rex* would present.

The trilobites were a huge and very successful group of arthropods that diversified into at least 4,000 species. All these species are extinct and none were ancestral to any surviving animal line. Several whole classes of Echinoderms have gone extinct, as have more than 10 orders of insects. Though insects are the most diverse animal class, with almost 1,000,000 known species, they were even more diverse back in the Permian Period, having lost in total species numbers since then. These examples are all from the animal kingdom (with which I am more familiar), but there are good examples of large extinctions in the plant kingdom as well. Cycads (Figure 8.1) are known and cultivated by a few plant fanciers, but today they are found naturally only here and there in the tropics. Back in the Jurassic, cycads were almost everywhere, making up roughly one out of every five land plant species. Now only a few species in 11 genera remain. At least 19 other cycad genera are extinct. If we had a decent fossil record for the Protoctista, Fungi, and Monera (which we don't) they too would undoubtedly show the same pattern of large-scale extinction of species, genera, families, orders, classes, and even phyla.

One of the earliest records of multicellular life is the Ediacaran fauna from around 600,000,000 years ago. When first discovered in 1947, most of the forms (Figure 8.2) were quickly assigned to modern animal groups. Since then, more careful studies have shown that though some of these organisms were probably early representatives of modern animal phyla, some were so unique that they probably represent distinct phyla which seem to have gone extinct. Slightly later fossils beds such as the famous Burgess Shale in Canada show few signs of life-forms similar

Figure 8.1 A cycad. Cycads form a primitive group of plants, which has dwindled over time to only a few surviving species. (*Photo by Leon Jernigan*)

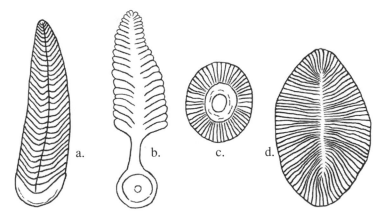

Figure 8.2 A few examples of the Ediacaran fauna from just under 600,000,000 years old: (a) *Spriggina*—probably a worm, (b) *Charniodiscus*—probably a Cnidarian colony, (c) *Cyclomedusa*—either a jellyfish or a solitary polyp Cnidarian, (d) *Dickinsonia*—probably a worm. (*Figure by author*)

Table 8.1 The Five Known Major Extinction Events of the Last 500,000,000 Years

1. 440,000,000 years ago: End of the Ordovician; wiped out many marine groups; about 25% of families and 60% of genera of marine invertebrates lost.
2. 370,000,000 years ago: End of the Devonian; 70% of marine animals and 19% of families lost.
3. 250,000,000 years ago: End of the Permian; most extensive single extinction event; in the marine environment over 50% of animal families, 80% of genera, and over 90% of species (including the last of the trilobites) were lost.
4. 210,000,000 years ago: End of the Triassic; 23% of families lost.
5. 65,000,000 years ago: End of the Cretaceous; 50% of genera lost, including all the dinosaurs and the ammonites; almost certainly due in large part to a massive asteroid which collided with the earth near Yucatan in Mexico.

to many of the Ediacaran forms. The specialists even argue as to what kingdom some of the Ediacaran forms represent. There are speculations that some may have been flattened mats of photosynthetic cells (algal-like), or sheets of cells which housed chemosynthetic bacteria, or they may have simply absorbed dissolved nutrients from the seawater. In short, these early multicellular organisms could have been a mix of strange forms, some of which correspond to nothing in today's roster of taxa.

Were these extinct groups failures? Of course not, how could we proclaim that over 99 percent of all the species that ever lived were failures? All species are destined to become extinct or to evolve into different species. There have been at least five large-scale "extinction events" which have evidently been due to planet-scale perturbations such as asteroids striking the earth, large-scale volcanic activity, continental collisions, rapid climatic changes, drastic sea-level changes, etc. (Table 8.1). The large numbers of species which were wiped out in these disastrous events were often the unlucky, not necessarily the poorly adapted.

Even a well-adapted species will become extinct if its niche is obliterated. Consider, if you can, the undoubtedly many diverse parasites that had specialized on/in dinosaurs. When the dinosaurs went extinct, so went the niche into which these well-adapted parasites had specialized. This brings up the point that when a species becomes extinct, it often carries several other species with it. This is yet another reason for the large percent of extinct biodiversity.

The five mass extinction events back through life's history, though responsible for wiping out indiscriminately (or nearly so) vast parcels of biodiversity, did set (or clear) the stage for the adaptative radiation of new species and new ecosystems which resulted in even greater temporal diversity of life forms. These five extinction events are only recognizable during the last 450,000,000 years, which is only one-seventh or so of the time life has been present on earth. Much before that, prokaryotes, simple protists, and soft-bodied multicellular forms were earth's only inhabitants. Surely, there were major extinctions events during this long earlier portion of life's history as well, but we have no obvious fossil record from which to detect such events among those earlier and poorly preserved organisms. Life today is the product of a long and complex set of historical events and evolutionary processes.

As some writers have pointed out, it is this long historical drama of change and extinction that gives a direction to time in the living realm. It has provided a time scale along which we can arrange many of the events and players in life's history—events such as the origin of life, the first eukaryotic cells, the five major extinctions, the first land animals, the first seed plants, etc. along with the major players such as the dinosaurs, cycads, and trilobites. Time is thus another dimension along which biodiversity varies, and varies immensely. Life has moved from a world of prokaryotes three billion years ago, to a world of emerging plant and animal forms some 600,000,000 years ago, to the first land animals some 410,000,000 years ago, to the world we see today with its wide span of biodiversity, which still includes a great diversity of the simpler ancestral forms. Each adaptation, speciation event, extinction event, shift of niche, etc. has affected the cast of characters composing life on earth in its successive historical epochs. Undoubtedly, the future of life holds many more unforeseen events and characters that will continue to expand the range of biodiversity along its long temporal span.

Though the millions of extinct forms are no longer living, they are certainly a part of biological science because they serve as examples for comparison, evidence of evolutionary patterns, indicators of evolution's speed and constancy (or lack thereof), and clues to modern-day species distributions and ranges. Only through our knowledge of extinct species do we recognize which modern groups are new, or ancient, or expanding, or declining, etc. We include these extinct forms in our modern taxonomic categories as plants, arthropods, vertebrates, etc. We even reconstruct as best we can the behaviors and ecologies of these extinct species. Evolution is said to be the cornerstone of biological science, so of course we are interested in the evolutionary history of our modern groups and the biodiversity of past ages.

9

Morphological Diversity

... from so simple a beginning endless forms most beautiful and wonderful have been, and are being, evolved.
—Charles Darwin, *The Origin of Species*, 1859

Morphology refers to the form and structure of an organism along with the form and structure of all recognizable "parts" which make up an organism. It includes the color, the size, the volume or mass, the shape, the density, the internal anatomy, the cell types, the number of distinct tissues and organs, etc. It might seem that since almost every species is typically distinct in their morphology (except in the case of sibling species—see Chapter 3), this chapter couldn't add much to an understanding of biodiversity. In fact morphological diversity occurs at several levels in the biological realm.

MORPHOLOGICAL DIVERSITY BELOW
THE LEVEL OF INDIVIDUALS

Can biomolecules be said to have morphology? Well, they certainly do vary greatly in size and shape, and they have different properties according to their size, shape, and charge characteristics. Each individual from bacteria to wolves produces thousands of distinct biomolecules. If we allow that proteins, complex carbohydrates, etc. have morphology (as biochemists do), there is certainly a great deal of morphological diversity at this level within each individual, and of course even more within and among species. A more detailed discussion of this level of diversity is given in Chapter 6.

In colonial and multicellular organisms, cells too can vary greatly within the same individual. Our nerve cells look drastically different

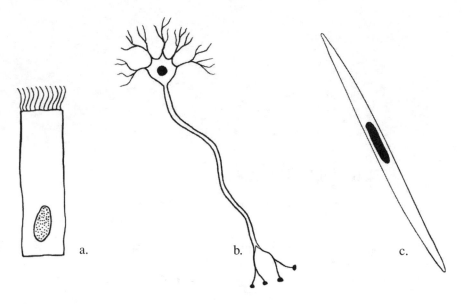

Figure 9.1 Examples of cellular diversity in vertebrates: (a) a ciliated epithelial cell, (b) a nerve cell, (c) a smooth muscle cell. (*Figure by author*)

from epithelial cells, which look very different from muscle cells (Figure 9.1). The ultrastructure of different cells in the same body can vary
as well. Gland cells typically have vast amounts of endoplasmic reticulum
as compared to a fat cell or a sensory neuron. In multicellular organisms,
cells are grouped into tissues which can be quite morphologically distinct. The discipline of histology concerns itself with these distinctions
of tissue morphology. Other particular body structures can vary significantly in morphology such as our vertebrae (Figure 9.2), or incisors
and canines, large intestines and small intestines, etc. Though these examples are often discussed as examples of "complexity," they are also
examples of biodiversity at organizational levels below the individual.

MORPHOLOGICAL DIVERSITY AT THE LEVEL OF THE
INDIVIDUAL AND SPECIES

Variation at the level of the individual is a topic not often addressed by
biologists. One can encounter this level of diversity in the area of animal
behavior where variation can occur in an individual's performance of
the same behavior from execution to execution, something that doesn't
apply in morphology. If a wasp's tibia has two stout dorsal bristles now,
it will have two dorsal bristles 10 minutes from now, but in the calling

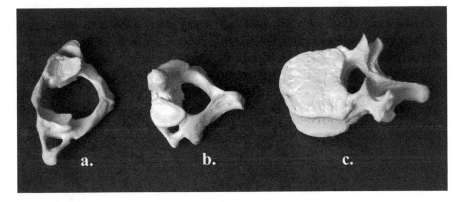

Figure 9.2 Some of the diversity of human vertebrae: (a) the atlas or first cervical vertebra, (b) the axis or second cervical vertebra, (c) a thoracic vertebra. (*Photo by Leon Jernigan*)

behavior of some insects, a male often delivers several calls (to a female) which, while all being characteristic of the species, will sometimes show measurable variation from one to the other. But behavior is not morphology, so how can there be morphological variation within an individual?

What about the effects of aging? Does age not often alter the morphology of some organisms? Certainly we know that size will often increase with age. When we meet old acquaintances whom we haven't seen in years, we often notice obvious changes in the person's appearance such as weight changes, graying hair, hair loss, skin changes, altered posture, etc. A person of either sex will show some obvious morphological changes during puberty. Some young plants produce leaves that don't look much like those of an older tree (Figure 9.3). Sockeye salmon undergo a radical alteration in morphology during their last weeks of life as they leave the ocean and swim upstream to their spawning streams (Figure 9.4). In many species, morphological changes with age are obvious, typical, predictable, and normal components of the life process.

What of those organisms that undergo complex life cycles? Each butterfly or moth transformed from a caterpillar into the winged adult, a truly radical metamorphosis in which almost all the morphology of the caterpillar is destroyed in the process of building up a new morphology of wings, new mouthparts, antennae, legs, reproductive structures, etc. all of which were lacking in the caterpillar stage. We are all aware that frogs undergo a similar, though less drastic transformation in their metamorphosis from tadpoles to adult frogs.

a,

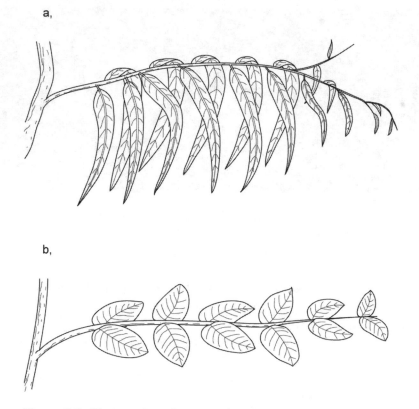

b,

Figure 9.3 Obvious size, shape, and arrangement differences between the mature leaves (a), and the young juvenile leaves (b) of *Eucalyptus globulus*. (*Figure by Jeff Dixon*)

Larval Echinoderms such as sea stars and sea urchins look absolutely nothing like the adult animals (see Figure 4.1), and the same goes for larval versus adult barnacles (Figure 9.5). A great many of the earth's species undergo such alterations and transformations in morphology. Some species, such as some parasitic flukes, have up to five morphologies within the same life cycle (see Figure 13.3). Textbooks stress, and biologists mainly concentrate, on the morphology of the "adult" or more obvious stage of the organism, whether plant or animal, but why? In the case of some insects such as mayflies, the aquatic larval stage makes up more than 99 percent of the life of the individual, with the short-lived adult only surviving for a matter of hours or at best 2–3 days (only long enough to mate, lay eggs, and die). The larval mayfly is perhaps where our emphasis should lie because they are obviously the more

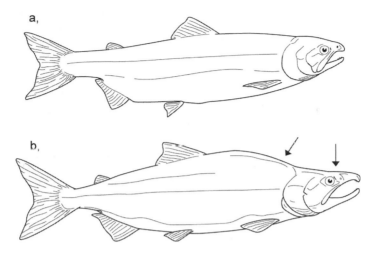

Figure 9.4 The relatively rapid and drastic metamorphosis that occurs in Sockeye salmon (*Oncorhynchus nerka*) as they move from their marine habitat (a) and start to swim upstream to spawn (b). The color also changes from mainly silver in the marine form to a brilliant red in the spawning individuals. (*Figure by Jeff Dixon*)

significant form from a temporal and ecological standpoint than the sexually mature adult.

Though in many of the marine invertebrates the larval stage can be much shorter-lived than the adult, the larvae can be equally or more importantly ecologically because they exist in incredible numbers, supplying food to other organisms in the complex marine food web. A life cycle doesn't really have a beginning or an end, and we overlook much of the earth's biodiversity if we concentrate on one stage, sometimes mistakenly referred to as the "final" stage, to the exclusion of the others. It is becoming more and more obvious that for many of the world's species we have too often placed unjustified emphasis on only one stage in a varied and complex life cycle. All stages are worthy of emphasis, study, and comparison, even if it does make organismal and ecological biology more complex than it already is. Life cycle diversity is discussed further in Chapter 13.

Some morphological variation is adaptive on a seasonal basis. Snowshoe hares have thick white winter coats, which help conceal them in snowy landscapes, but which molt in spring to reveal a new thinner brown coat that is more inconspicuous against spring, summer, and

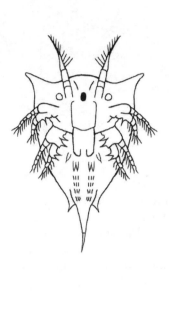
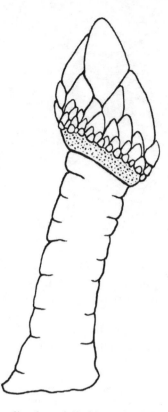

Figure 9.5 Comparison of the nauplius larval (left) and adult (right) morphology in a gooseneck barnacle. The nauplius larval is shown greatly enlarged. (*Figure by author*)

autumn landscapes. Males in some species of deer and elk grow antlers for part of the year, and then shed them in annual (or sometimes longer) cycles. Some male fish and birds develop their bright breeding colors only during the breeding season and look far less brilliant the rest of the year. Deciduous trees obviously have a different morphology in summer as opposed to winter when they loose their leaves, flowers, etc. Most species of mushroom-forming fungi only grow the above-ground mushroom for a few weeks out of the year, whereas the underground hyphae survive throughout the year.

Many species are known to be polymorphic, not due to age, or season, or life cycle stage, but to some other factor. Species can be polymorphic in many ways. Many species are at least dimorphic according to sex, that is, males look noticeably different from females, sometimes so much so that they could in ignorance be taken for separate species. Most eusocial insects are polymorphic according to caste such that queens look different from workers, which look different from soldiers. Other amazing cases of species polymorphism exist in certain species of tropical butterflies which contain up to three or four coexisting "morphs" with all, or all but one, being a Batesian mimic of another species of local butterfly which is toxic or distasteful to predators.

In Pacific salmon, a rather unique polymorphism has been elucidated among the males of the species. Most newly developed males leave their freshwater streams to enter the ocean for several years of growth and development. A few males, referred to as "precocious parr," never go to sea, rather they remain in the freshwater streams where they hatched to grow into small, but reproductively adult males. Parr are too small to compete directly with the larger and more aggressive typical males, but they can sometimes "sneak" their sperm onto freshly laid batches of eggs just as a typical male releases his sperm. Some of these eggs are being fertilized by parr sperm, otherwise, parr would not exist.

There is yet a third type of male among the Pacific salmon. This third type is termed a "jack." Jacks do go to the ocean, but they return to their rivers only after a year. Jacks are also smaller than the males who spend years maturing at sea. They are apparently successful in swimming back upstream to spawn, but like the parr, they are poor competitors with the larger "normal" males. Like the parr, jacks will attempt to "sneak" some of their sperm in with that of larger males as or slightly before the larger male releases his own. Such morphologically distinct "sneaky males" are also known to occur in other species such as the bluegill sunfish.

A species can also be polymorphic due to racial development. In less than 200 years house sparrows have evolved distinct races that look different in different parts of North America. Many species have evolved geographical races that are morphologically distinct (at least on average) from one another. People are still rightly or wrongly assigned to racial groups, which were certainly more distinct and disjunct prior to widespread access to global travel.

When it comes right down to it, morphological diversity is widespread within almost any sexually reproducing species. One of the facts of nature upon which Darwin founded his principle of natural selection was that sexually reproducing species give rise to variant offspring. Since Darwin was not well aware of molecular and other unobvious variations, he was mainly referring to morphological variations. Look around on any crowded street and try to find two people who are morphologically indistinguishable. You will be out of luck unless there happen to be identical twins among the crowd. Though other animals look to us to be

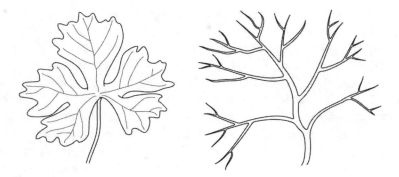

Figure 9.6 Leaves of the water buttercup (*Ranunculus aquatilis*) according to environment. Leaves that form in the air (left) are filled out with more surface area to collecting sunlight for photosynthesis, while leaves that form underwater (right) develop a finely divided morphology which is more efficient at collecting dissolved oxygen, and also provides less resistance to water movements. Both types of leaves may be present on the same plant if it grows in a situation where it is only partially submerged. (*Figure by Jeff Dixon*)

virtually identical, they are not. Several years ago a National Geographic article reported on a study of rhinos in Africa. The biologist carrying out the study (Hans Kruuk) was able to recognize all the rhinos in his study area individually by morphological differences in their ears, horns, size, skin folds, etc.

Environmental factors can cause intraspecific morphological variation as well. In some animals, varied diets can result in obvious morphological differences, even when individuals are genetically identical. Two identical twins could easily be of different heights and weights if they had been raised from their early years on drastically different diets. Their skin tone and texture could also vary with differential exposure to sunlight.

In plants there are numerous cases where different parts of the same plant, or different individuals of the same species, show different morphologies because of environmental factors (Figure 9.6). Parasites and disease can have morphological effects on their hosts. Though humans tend to view the presence of such parasites as "abnormal," in nature a sizable proportion of a population may be affected and thus show the morphological "abnormalities." In large areas of the Southern United States, hackberry trees can be recognized in part by the presence of certain distinctive galls (growths stimulated by insect eggs and larvae) on their leaves, and in those areas it can be almost impossible to find a hackberry tree without them.

MORPHOLOGICAL DIVERSITY BETWEEN SPECIES AND HIGHER TAXA

Morphological diversity between species is almost always present except in the case of sibling species, which are likely much more numerous that most biologists suspect. Again, there are around 2,000,000 described species, and probably at least 8–10 million actually in existence. Most of these do vary from one another in at least minor morphological features such as the shape of the genitalia between two closely related cricket species. Of course when we compare crickets to mushrooms, or bacteria, or pine trees, the morphological differences are so great that we don't even have an obvious basis for comparison. There is little that can be compared between crickets and mushrooms except at the cellular and molecular level because their tissues and body parts simply don't correspond in any obvious way. Crickets and beetles could be compared in several meaningful ways since they are both insects and have systems and body parts that do correspond in most cases. There could be direct quantitative comparisons between the legs, wings, eyes, antennae, mouthparts, respiratory systems, nervous systems, glands, etc. of crickets and beetles, but since mushrooms have none of these features, comparisons are less possible, except to point out presence/absence differences.

Quite clearly, there are obvious morphological differences between many of the earth's species, some great and some small. There is no need to belabor the obvious morphological differences that exist between species of cacti and ants, bread molds and *Paramecium*, ferns and junipers, or flounders and angelfish. A different aspect of morphological diversity arises when we consider the higher taxa. Millipedes make up the class Diplopoda, while mammals make up the class Mammalia. The number of known millipede species is around 10,000, while mammals number less than half that number—around 4,800. So, millipedes contain more diversity than mammals—right? If we are talking only about species diversity, then millipedes are clearly more diverse. But, in terms of morphological diversity (technically termed *disparity*), most biologists would argue that mammals display a much greater range of morphological diversity than do millipedes. Though there are fewer mammal species, mammals have diversified into several distinct and very different body plans. Bats are drastically divergent from whales in their morphology, just as bears are drastically different from mice. We humans are morphologically quite divergent from most other mammals.

Both in size and in body construction, mammals include a very diverse assemblage of body plans and sizes, whereas millipedes all look basically like, well—millipedes. Sure, each millipede species is in some way morphologically distinct from the others, but different in minor ways (size, number of body segments, number of legs, minor shape differences, color, etc.) as compared to the larger differences between a camel and an anteater, a human and an otter, a flying squirrel and a tiger, or a porpoise and a elephant. Also, mammals are more completely known with almost certainly 80—90 percent of species having been described. Millipedes on the other hand are not nearly as well known. The 10,000 known millipede species most probably represent far less than 50 percent of actual species.

Some groups of higher taxa clearly display more or less morphological diversity than do others, even of the same taxonomic level, as in the example above. Though there is no clear rule or formal measure of the degree of such morphological divergences, it is nevertheless obvious that biodiversity is in part composed of such disparity differences that do not vary in close conjunction with our classification system. Monerans, which make up two of the three domains of life, and which have had a much greater span of time than other life-forms to diverge into different taxa, still have such simple morphologies that there is not much that could be drastically different between them except at the level of their molecules and metabolisms. We really don't know much yet about how diverse Monerans really are, but the guess is that they are extremely diverse at the level of species. Again though, it is clearly arguable that some more recent and smaller groups such as the Cnidarians (jellyfish, hydra, corals, sea fans, sea plumes, etc.) display more morphological biodiversity than the entire kingdom (or kingdoms) of Monerans. To the extent that multicellular organisms are more complex than prokaryotic unicells, they have that much more potential for morphological diversity because there is simply more morphology "available" for adaptive divergence.

Consider hair. If a species has hair, as most mammals do, then there is great potential for morphological diversity that is not present in animals lacking hair (though they may have another feature with great diversity potential such a feathers, shells, flowers, etc.). Within the mammals there exists a great diversity of hair morphology. Much of a person's body hair is vestigial "fuzz," especially in women. The hairs on the head are entirely different in texture, length, and usually in color. The underarm and pubic hairs are also distinct in length, texture and color, as are

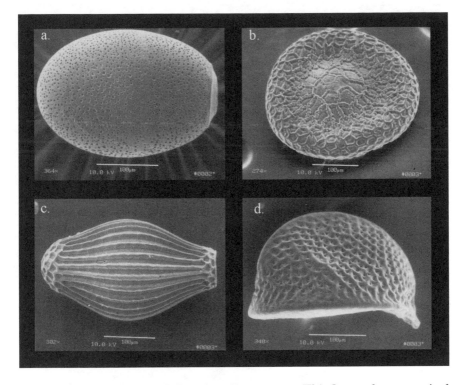

Figure 9.7 Morphological diversity of insect eggs. This figure shows a typical egg for each of four species of stoneflies (order Plecoptera). (a) *Paraperla frontalis*, (b) *Yoraperla nigrisoma*, (c) *Neoperla stewarti*, and (d) *Diploperla duplicata*. (*Scanning electron micrographs courtesy of Dr. Bill Stark, Mississippi College*)

eyelashes. Some people have distinct "spots" on their heads which sprout gray hair while the rest is brown or black. Most of our hair grows only to a certain length while the hair of our heads and beard will grow to almost any length allowed (a puzzling fact—how is this adaptive?).

The long sensory whiskers of cats are obviously modified hairs, as are the mane and most of what we call the tail in horses. In lions, the male's mane contributes obviously to sexual dimorphism. The stiff quills of a porcupine are modified hairs. Even the horn of a rhinoceros is actually a mass of tightly compressed modified hair. Hair can vary between taxa in color, color pattern, length, thickness, etc. Even the loss of hair as has occurred in whales and dolphins contributes to diversity because it creates such a distinct contrast to other mammals.

Morphological diversity exists not only in the body of the organism itself, it can also show up in many of the "products" of organisms. The pollen, fruits, and seeds of many plant species are distinctive, often to the species level. The eggs of many animal species are likewise unique (Figure 9.7). Many bird species can be identified by the nest that they build, or by their bowers in the case of the bowerbirds.

To sum up, morphological diversity exists at several levels below the individual such as that of genes, molecules, cells, and tissues. Morphology can vary within individuals as in age- or environmental-related phenomenon. Morphology is certainly diverse at the species level as in sexual dimorphism, caste polymorphism, and the individual variation that Darwin recognized. Finally morphology varies most obviously between species (excluding sibling species) and within and between the higher taxa, though the degree of morphological variation at higher levels is unpredictable and variable from group to group.

10

Metabolic Diversity

Comparative biochemical studies reveal that the details of protozoan metabolism are as varied as the details of protozoan sex.
—S. Roberts & J. Janovy, Jr. *Foundations of Parasitology*, 6th ed., 2000

... through the invention of a spectacular array of different metabolic and physiological capabilities, microbes evolved to exploit the multitude of environments and microhabitats presented by the abiotic world.
—Paul V. Dunlap, Microbial Diversity, in *Encyclopedia of Biodiversity*, Vol. 4, 2001

Another of the basic characteristics of living organisms mentioned back in Chapter 1 is metabolism. To be alive, some metabolic activity must be present. Many people claim to have died and returned to life with descriptions of the "afterlife." Though a heart may stop beating, life has not really stopped. Most of the cells in our body can continue to survive for a few minutes without a heartbeat, and some can survive much longer with continued metabolic activity. Muscle cells can survive for more than 2 hours after the heart stops. Even nerve cells can survive for several minutes without blood circulation, depending on the temperature. Colder temperatures seem to favor longer survival times, as when someone is revived after up to 20 minutes or more submerged in frigid water. We die by degrees, not at one precise moment. For humans, if "you" are truly dead, that's it, even if some are resuscitated back to active life after near brushes with the real thing.

Amazingly, there are a few creatures that can move from metabolically active life into a state that appears to be death, and back again. Some bacterial spores, some plant seeds, and some dried small animals such as water bears and rotifers can remain viable without any measurable metabolic activity. This state of "suspended animation" (to take a term from some old science fiction movies) can in some cases last for many years. The resting state ceases when the organism reanimates or continues its metabolism, growth, and development due to the reintroduction of favorable environmental conditions, one of which is usually the increase or addition of water. Life is a process—in large part a metabolic process. In these "suspended" organisms the process appears to have stopped completely, yet it can return to the same seed, spore, or water bear when conditions favorable to the process return. Such organisms are the exception, because in the majority of the earth's organisms metabolic processes must be continuous at some level for life to continue. Remember that a bear's metabolism slows during hibernation, but it does not stop.

Metabolism has a few common themes and threads among the variety of earthly life, but there are many major and minor variations unique to certain groups (the old unity and diversity principle once again). A current reawakening of interest in the study and discovery of various microbial metabolisms is directing attention to the truly broad spectrum of metabolic lifestyles present among our fellow creatures. Though most metabolic diversity seems to exist among the prokaryotic groups, there are variations within the eukaryotes as well, many of which are undoubtedly still awaiting discovery and elucidation, especially among the Protoctista. We will only have time in a single chapter to touch on a few of these major metabolic processes and some of their variations, but these should serve as a revealing introduction to the vastness and complexity of metabolic diversity.

PHOTOSYNTHESIS

This important metabolic process is taught to most students once or twice in their precollegiate studies, and typically once more (depending on their major) at the university level. Many organisms from three of life's traditional kingdoms (prokaryotes, protists, and plants) are capable of photosynthesis. Photosynthesis as we commonly think of it originated in certain bacteria, the cyanobacteria, and was passed to some ancestral protists as described by the endosymbiotic theory of eukaryotic cell origins; that is, the chloroplasts in eukaryotic cells are descended from free-living cyanobacteria which became endosymbionts of other cells, making those cells photosynthetic. Plants evolved directly from one group of photosynthetic protists, the green algae.

Photosynthesis is almost always taught from the standpoint of aerobic noncyclic photosynthesis (shown below), which is by far the most common type, occurring in plants and most photosynthetic protists and bacteria. Perhaps this is wise, since covering even this standard form of photosynthesis is quite challenging, and introducing the student to any variations might lead to confusion. Unfortunately though, most students, even those in the sciences, end up unaware of the variations that are present in some groups of photosynthetic organisms.

$$CO_2 + H_2O + SUNLIGHT \rightarrow (CH_2O)_6 + O_2 + H_2O$$

The water on the input side of this generalized equation (this equation is not balanced for simplicity sake) supplies both electrons and protons from its hydrogen atoms. They end up as part of the sugar or $(CH_2O)_6$ molecule while the remaining oxygen escapes as the product oxygen on the right side of the equation. A relatively minor variation of "standard photosynthesis" is referred to as C_4 photosynthesis. The generalized equation would look the same as that shown here, but additional intermediate steps occur which allow plants that live in hot dry climates to carry out photosynthesis more efficiently. The details of this variation are too complex for our discussion here, but they can be found in any general botany text. A more major variation of photosynthesis occurs in some photosynthetic bacteria, specifically the so-called green bacteria and the purple bacteria, which use something other than water as their source of electrons and protons. One such variation is to use hydrogen sulfide (H_2S) rather than water. The generalized equation then becomes:

$$CO_2 + H_2S + Sunlight \rightarrow (CH_2O)_6 + S_2 + H_2O$$

Some species can use substances other than hydrogen sulfide in a similar way. There are of course other differences in the mechanisms and steps of this form of photosynthesis as compared to that of plants.

CHEMOSYNTHESIS

Until the late 1970s, most textbooks continued to say things like "all the organic material on earth originated from photosynthetic producers." They strongly implied that all life on earth was dependent, at least indirectly, on the organic molecules produced by photosynthesis. It was an oversimplification then, and today such a statement should clearly be recognized as false. We have now discovered several deep-sea, deep-earth, and a few cave ecosystems (see Chapter 7) which are fueled not by photosynthesis but by a variety of producer prokaryotes that use chemosynthesis to produce their organic molecules.

The prefix "chemo" tells us that chemicals are the energy source here rather than light (photo), but the synthesis ending says that energy-rich molecules (carbohydrates) are being synthesized, just as in photosynthesis. Only some of the prokaryotes carry out chemosynthesis. Though chemosynthesis was known since the early 1900s, it was not until 1977 that we discovered that whole communities and ecosystems could be founded on chemosynthetic processes. This was the year the first deep-sea hydrothermal vent community was discovered in the deep Pacific off the coast of Equador, and many other chemosynthetic communities have been discovered since then (see Chapter 7).

So, we now know that there is a way of manufacturing usable chemical energy other than photosynthesis. This is diversity in and of itself, but the real diversity is in the variety of chemosynthetic pathways. First let's look at a representation of one common form of chemosynthesis that is known to occur in some of the hydrothermal vent communities.

$$H_2S + O_2 \rightarrow S + H_2O$$
$$\downarrow$$
$$\text{Energy (ATP \& NADPH)}$$
$$\downarrow$$
$$H_2O + CO_2 \rightarrow \text{Calvin Cycle}$$
$$\downarrow$$
$$\text{Carbohydrate} + H_2O$$

I know this looks quiet complicated, but this is actually a shorthand version of the more complex process. The first line of the process is the "chemo" part. Hydrogen sulfide (H_2S) is oxidized into elemental sulfur and the oxygen is reduced to water. This process releases energy which is channeled into the buildup of molecules of ATP and a reduced substance abbreviated NADPH. These same two energy-rich molecules are produced in photosynthesis by the harnessing of light energy. In short, chemical energy replaces light energy in chemosynthesis. The lower half of the process is essentially the same as for photosynthesis. The Calvin Cycle is the later stage in photosynthesis which generates carbohydrate units. Other chemosynthetic bacteria utilize other energy-yielding reactions of inorganic substrates as their energy source. Some of these are

$$NH_3 + O_2 \rightarrow HNO_2 + H_2O$$

$$Fe^{2+} + O_2 + H_2O \rightarrow Fe^{3+} + OH^-$$

$$S + O_2 + H_2O \rightarrow H_2SO_4$$

$$H_2 + O_2 \rightarrow H_2O$$

There are many others, and these chemosynthetic bacteria also vary in the details of how the energy is harnessed to manufacture carbohydrates. In part due to chemosythetic bacteria, we can safely say that the majority of metabolic biodiversity exists in the realm of the prokaryotes. Since they are the oldest and most adaptable forms of life on earth, this only makes sense.

FERMENATION RESPIRATIONS

Once stable organic chemical energy is formed by some variation of photosynthesis or chemosynthesis, organisms must then extract that energy in the form of ATP, GTP, or other intermediate molecules to run the chemical processes of their cells (refer back to characteristic #10. in Chapter 1). The term respiration broadly refers to the release of energy from the oxidation of fuel molecules. Oxidation refers to the removal of electrons (and often whole hydrogen atoms) from substrate (fuel) molecules. The fuel molecules are typically sugars, amino acids, lipids, etc. One of the simpler forms of respiration is termed "fermentation." Fermentation does not require oxygen, yields few ATP molecules per fuel molecule, and is used by a variety of unicellular and multicellular organisms.

One form of fermentation carried out by yeasts and several other microorganisms is alcohol fermentation. This is the process that has long been harnessed by humans in the making of alcoholic beverages. The generalized equation for the alcoholic fermentation of glucose is

$$(CH_2O)_6 \rightarrow Ethanol + CO_2 + Energy \ (heat \ \& \ ATP)$$

Some other bacteria and fungi utilize a different form of fermentation called lactic acid fermentation. Some of these have been harnessed in the production of dairy products such as cheese, yogurt, and buttermilk.

Interestingly, our own skeletal muscles can switch from aerobic cellular respiration (illustrated later) to lactic acid fermentation when we are exercising strenuously and oxygen is not being delivered fast enough to the cells to keep up with the muscle's demands for ATP molecules. The generalized equation for this process is

$$(CH_2O)_6 \rightarrow \text{Lactic Acid} + \text{Energy (heat \& ATP)}$$

Another variation on fermentation is known as formic acid fermentation. Here, formic acid is an intermediate in the process, rather than an end product. Many species of the common genus *Escherichia* (as in *E. coli*) utilize this process:

$$(CH_2O)_6 \rightarrow CO_2 + H_2 + \text{Energy (heat \& ATP)}$$

Yet another variation is butanediol fermentaion, utilized by other bacteria.

$$(CH_2O)_6 \rightarrow CO_2 + \text{Butanediol} + \text{Energy (heat \& ATP)}$$

That makes four variations, and yes—there are others, but these four should illustrate the diversity within this class of metabolism well enough.

AEROBIC RESPIRATION

We often think of respiration as the breathing process, but more correctly the term applies to the metabolic process that involves the oxygen (which we breathed in) that occurs in the cells of our body. It is sometimes also called *aerobic cellular respiration*, to distinguish it from the breathing process. This is another of those processes that any high school graduate has been exposed to at least once, with most college graduates having had another exposure. Biology graduates most likely cover it twice in differing levels of detail during their undergraduate years. We were usually told that this process is how the vast majority of the earth's species extract usable chemical energy (ATP molecules) from the energy stored in carbohydrates, lipids, and amino acids. This is an arguable point, because we now know that a great number and mass of the earth's organisms use processes which range from similar, to drastically different from standard aerobic cellular respiration. Still, the process is an important one that is widespread in most of the multicellular life-forms we are familiar with.

The process is usually described in terms of the aerobic respiration of a glucose molecule. Glucose and other simple sugars are probably the most commonly respired substrates. The generalized equation, which summarizes a great many separate steps, is as follows:

$$(CH_2O)_6 + O_2 + H_2O \rightarrow CO_2 + H_2O + Energy \; (heat \& ATP)$$

This form of aerobic respiration is what you are doing right now, and it is much more efficient in terms of producing ATP molecules than any of the anaerobic fermentation processes. The input oxygen is reduced with the electrons and protons from both the sugar molecule and the input water to become the output water on the right side of the equation. The substrate (sugar) is oxidized to release the energy, some of which is incorporated into the newly formed ATP molecules. Most multicellular organisms that require oxygen are utilizing this process, and this includes plants, as well as most fungi and animals. Depending on the organism, and even the tissue, either 36 or 38 ATP molecules are gained in the process (Chapter 4 explains these two different outcomes). Again, the previous generalized equation (like all those in this chapter) is a great oversimplification of the entire process, which involves over 50 separate chemical reactions.

Trypanosomes, the parasitic protists that cause African sleeping sickness, typically utilize oxygen in their respiration of glucose, but their form of aerobic respiration is a unique process which only partially catabolizes the glucose and yields only two ATP molecules. Water is again formed as some hydrogens from the glucose are passed eventually to the oxygen molecules.

$$(CH_2O)_6 + O_2 \rightarrow Pyruvic \; Acid + H_2O + Energy \; (heat \& ATP)$$

Instead of using oxygen as the electron (and possibly proton) acceptor, some organisms carry out a type of respiration utilizing a different electron acceptor. Some bacteria use sulfate (SO_4) as their electron acceptor. Their respiration process would be represented by this generalized equation:

$$(CH_2O)_6 + SO_4 + H_2O \rightarrow CO_2 + H_2S + H_2O + Energy \; (heat \& ATP)$$

Some of the bacteria that use this process live in anaerobic muds. Those who have walked in such mud will remember the repulsive smell of the output hydrogen sulfide (H_2S—which smells like rotten eggs) that is released when you pull your feet out of these anaerobic layers. There are still other variations on this basic process depending on which molecule is utilized as the electron acceptor.

These are but a few of the major metabolic processes and variations that occur in the diversity of organisms now living on our planet. We have not covered here other common metabolic processes such as the conversion of carbohydrates to lipids, the synthesis of polysaccharides such as starch or glycogen, the synthesis of proteins, etc. All these and more have their variations among the various species of our planet. Still more metabolic diversity exists in

- Metabolic rate: Ectotherms such as reptiles generally have lower metabolic rates than endotherms such as birds and mammals. Hibernating mammals have lower metabolic rates than nonhibernating mammals. Bacteria that live in polar ice and in tiny cracks in rocks deep in the earth may have the lowest metabolic rates known, while some of the bacteria that live in our colons have exceptionally high metabolic and growth rates.

- Digestion: Fungi digest their food outside their cells and "bodies" by secreting digestive enzymes onto the food they are in contact with, and then absorbing the digested nutrients into their cells. Animals typically ingest food into a digestive system where enzymes do their work and nutrients are absorbed. Some animals digest macromolecules in a digestive cavity (stomach, intestines), some digest them in food vacuoles inside cells (sponges), and some do both (many of the invertebrates). Tapeworms have lost their digestive systems since they live in their host intestines where host enzymes do the necessary digesting. They simply absorb some of the nutrients intended for the host.

- Use of fuel molecules: While many animals regularly utilize lipids as alternate fuel molecules in respiration pathways, some parasitic animals like flukes and tapeworms use almost none. Animals often use amino acids as fuel molecules while plants almost never do.

- Excretion of nitrogenous metabolic wastes: When animals utilize amino acids as fuel molecules, a molecule of ammonia (NH_3) is typically released. Since ammonia is quite toxic, it must be excreted or converted to less toxic products. Most aquatic invertebrates excrete the ammonia itself, often by simply letting it diffuse out of their bodies and into the surrounding water. Some animals (including humans) convert some or most of their ammonia into urea. This significantly reduces the toxicity until the waste can be excreted. Insects, terrestrial snail, birds, and reptiles convert most of their ammonia into uric acid. Uric acid is far less toxic than ammonia and can be tolerated in relatively high levels until excretion. Spiders and scorpions convert their ammonia into guanine, which is less toxic still. There are a few other excretory products such as trimethylamine and creatine which can occur in combination with those discussed previously, depending on the species.

Metabolic pathways are very difficult processes to elucidate, and most of the earth's species have not even been investigated as to what new metabolic processes, or variations on known processes, they might harbor. Prokaryotes are only beginning to be well researched metabolically, and the eukaryotic protists are still little known from the metabolic perspective. For those brave enough to venture into this maze of metabolic pathways, there is a bewildering plethora of both known, and as yet unknown, metabolic diversity to be appreciated.

11

Sensory Diversity

There are a variety of sources of information about the world around us from which our species has been largely or wholly excluded.
—James L. Gould, *Ethology*, 1982

Each species moves, so to speak, in a particular sensory world from which other species are excluded in part or in toto.
—Francois Jacob, *Of Flies, Mice, and Men*, 1998

As mentioned back in Chapter 1, irritability is one of the basic characteristics of all life-forms. In order to respond to a stimulus and show the property of irritability, an organism must be able to sense that stimulus in some way. Much is known about sensation in animals, but sensation is not such an obvious topic in discussions of plants, fungi, protists, etc. A single chapter can only cover a sampling of the diverse sensory channels and methods used by various organisms in their attempts to gain important information about their environment through "the senses." Undoubtedly many variations, and possibly even some sensory channels, are still unknown.

We might begin this topic by addressing the old idea that humans have five senses, those being sight, hearing, smell, taste, and touch. To be brief, this notion is dead wrong. We clearly have more than five senses. It depends on where you draw the line on what constitutes a separate mode of sensation as to how many senses we humans possess. We certainly have a sense of balance, which is handled by structures associated with our inner ears. With this sense we can detect the force and direction of gravity, our orientation in space, and the force of acceleration. We also have a sense called proprioception, which is handled by a great number of mechanical sensors arrayed throughout the body, especially near joints and in muscles. Proprioception is the sense that allows us to know how our body and its parts are positioned in space, and in relation to one another (is our right arm straight or bent, is our head tilted or in line with our spine, is our left leg beside, in front of, or behind our right leg, etc.) Though we are usually not especially conscious of these sensations, we could neither walk, swim, climb stairs, type, play a guitar, nor do any complex activity without a continual stream of proprioception information coming from our body to our brain to inform the brain of where our arms, legs, hands, fingers, etc. are currently positioned in relation to one another. When a professional gymnast performs a complex floor routine, proprioception is far more important to the gymnast than the sense of sight.

We have a temperature sense which allows us to judge relative temperatures, as when we leave the warmth of our house on a winter's day. The temperature sense is not touch, which depends on pressure changes at the skin's surface, it is something quite different. Likewise pain is a unique sense that may or may not, depending on the circumstances, be associated with touch. It can be internal or external pain, and there are even different pain modes such as burning pain, sharp or stabbing pain, or throbbing pain, and what exactly is the pain of longing, or of a "broken heart," which can seem quite real and painful in their own unique way?

Do we have a sense of time? Though all waits in doctor's offices seem to take forever, some waits are obviously shorter than others, even when we don't have a watch or clock to confirm it. Surely all people have some abilities when judging and measuring relatively short time intervals such as a second, a minute, or an hour. We at least seem to have a sense of humidity, as detected when one walks into a room containing an indoor swimming pool or greenhouse, though officially this one has not been confirmed.

Are there others? More questionable and undocumented is the sense of "direction," which some people seem to have, while others don't. Much of this "sense" undoubtedly comes from simply having good spatial memory and the ability to "picture" in the mind the relative spatial distances and relationships of environmental landmarks and features. People also speak of a sense of rhythm in relation to music. Some people can easily find the beat or rhythm in even complex music such as jazz, whereas others can't seem to do so in even the simplest song. When it comes to abilities such as direction and rhythm, which many people "have" and many people do not, we are probably not dealing with real physical senses, but particular mental abilities, whether partially innate or learned. Though brains can reason and process certain information, the information itself arrives from the senses, such as hearing in the case of "rhythm." The brain itself is not a sensory organ, only the interpreter of the sensory input.

Any evolved sense with specialized outlying sensors and receiving centers in the nervous system should be present in all people, not only in some fraction of the population as seems to be the case for direction and rhythm. The supreme example of something that is almost certainly not a sense is "extrasensory perception" or ESP. Since only a very small percent of people claim to have this "sense," it is not a sense in the same way as sight, taste, balance, etc. There is no conclusive scientific evidence to show that ESP is even a real phenomenon. But even if we toss out direction, rhythm, and ESP, we still have an impressive number of senses for one species, and it illustrates well that sensory diversity can exist within a single species.

Another important point is that within the major senses, there can be variation among individuals. Some people can taste a chemical called phenylthiocarbamide (PTC for short), while other people cannot. This is a genetically controlled trait wherein some people manufacture a receptor for PTC, and some people do not. Some variations may be more complex than the case of tasting or not tasting a substance. It is at least possible that some people taste chocolate somewhat differently from other people due to more complex variations and interactions of taste receptors in those different individuals. Though two people can both agree that they love the taste of lemon pie, there is no way to prove that the two people are sensing this taste in exactly equivalent ways. Another example of individual variations within a major sense would be the several forms of color blindness, which each affects only a small percent of the population.

Not all animals are as diverse in their sensory abilities as we appear to be. Sponges certainly lack hearing and sight. It seems likely that they have no sense of pain, and being sessile and unable to move or withdraw, their sense of touch must be relatively weak to absent over much of their surface. Sponges have no nerves or brains, so their sensory world is undoubtedly extremely limited.

Ticks are more complex animals than sponges, but their sensory world too is quite limited when compared to ours. For those ticks that specialize on mammal hosts, locating a host requires only two simple stimuli. One of those is a chemical called butyric acid, a chemical common to all mammals, and the other is heat, which all mammals generate. Both cues aren't even registered at the same time. The chemical is sensed first, which causes the tick to drop or clamber from its perch in the vegetation, and then heat becomes the sole stimulus in locating the host. Having sensors for these two stimuli insure that the tick will likely locate an appropriate host.

After locating the host, the tick drinks blood, but apparently doesn't even taste it. Experiments have shown that any nontoxic liquid of the appropriate temperature will be imbibed. Ticks can wait a long time before a host wanders by. Some ticks apparently sit motionless for several years before the one or two sensory cues stirs the tick from its monotonous wait. Ticks do vaguely sense light, which aids in their positioning in the vegetation, and they sense one another through touch and chemical signals at the time of mating, but overall, their sensory world is one of extreme poverty when compared to ours.

Some animals that possess our same basic senses have abilities far beyond ours. Our sense of smell, though good compared to some animals, is quite feeble when compared to that of a dog. Dogs can smell many more compounds, and in concentrations far lower, than any human. Some insects can see ultraviolet and polarized light, which we cannot. Dogs and bats can hear sounds of very high frequencies (ultrasound) that are beyond the range of any human. Elephants and whales can hear sound of very low frequencies (infrasound), which again lies outside our range of detection. Pit vipers have specialized heat detectors on their snouts termed "pits" which allow them to "see" the radiant energy (heat) of their warm-blooded (endothermic) prey with great accuracy out to several feet. This feat is something the temperature receptors of your own skin could never do.

Even to say that another species has eyes does not say that that species can "see" in the same sense that we do. Planaria flatworms have "eye-spots," which are photoreceptors, but they only seem to sense light in terms of its direction and degree of brightness (Planaria avoid lighted areas where they might be spotted by visually orienting fish and other predators). Planaria almost certainly do not see anything like the multi-colored and detailed images our eyes (and brain) are capable of sensing, nor do many of the other "eyed" animals. Some animals can detect color, while others cannot (determined from conditioning behavioral experiments). Some can form sharp images of the shape and outline of objects in the environment, while others cannot. One visual aspect which most animals can detect is motion. Motion is very often important input because it may signal an approaching predator, a potential prey item, a potential mate, a caregiver, etc.

Some animals can even see *faster* than we can—that's right, faster. When light enters the receptor cells in our retina known as rods, the light causes a special pigment called rhodopsin to break into two parts; the brighter the light, the more rhodopsin that gets broken down. While the rhodopsin is broken down, messages of light sensation are sent to the brain's visual cortex and other areas where we then consciously sense the light. Over a matter of a few seconds, the two subunits of the rhodopsin are rejoined to renew the rhodopsin in the rods. When we pose for a flash photo, the very bright light of the flash breaks down a large portion of our rhodopsin, and it takes several seconds for that rhodopsin to be renewed. That is why we continue to see for a while the bright spot on our retina where the flash was focused.

While all that rhodopsin is being regenerated, we are effectively blind in that area of the retina. In a similar way, normal light levels and visual input "incapacitate" the rods for briefer periods of time, in effect holding the image on the retina after it has actually moved or vanished in the environment. This is why a fast-pitched ball appears as an elongated blur, or why a flashlight moved rapidly in the night seems to have a tail. The best example of this phenomenon is motion picture film, which is run at 48 frames per second. We actually see 24 still photos (each shown twice in succession) which are flashed on the screen each second. Each projected frame is separated by a totally black screen between the stills. We don't detect the reality of what is happening on the screen because our rhodopsins continue to signal the brain of the previous still during the intervening black screen until the next slightly different still is flashed in the same place. We end up seeing a continuous moving picture because of the lag time required for our rhodopsins to reform.

The time required for our eyes (actually our rhodopsins) to reset is termed our *flicker-fusion frequency*. Some animals, in particular some of the flying insects like the common housefly, have much higher flicker-fusion frequencies than we do. Without a long explanation, this simply means that their rhodopsins are different from ours, and they can reset in a much shorter period of time. The image on their eyes/brains lasts for a shorter period of time, so in the case of motion pictures, they could see the 24 stills separated by the black-screen intervals. More importantly for them, they could see more clearly (with less blur) any rapidly moving objects in space, which might be an approaching flyswatter, a bird or insect predator, a rival male, or a potential mate. Some flies chase each other in flight and some even couple and mate in flight. Seeing accurately in rapid time is essential for such fast-moving animals. See Table 11.1 for a comparison of the *various* flicker-fusion frequencies for a few animals. This diversity of flicker-fusion frequencies illustrates yet another example of biodiversity in a single sense.

Table 11.1 A Few Flicker-Fusion
Frequencies (FFF) for Various Animals

Animals	FFF (Hz)
Humans	60
Pigeon	120
Gecko	20
Goldfish	16
Cricket	45
Housefly or bumblebee	300

Note: Within a species, these numbers can
vary somewhat between individuals, and at difference light levels. (Compiled from several
sources.)

When we extend our survey of sensory abilities to all the species on
earth, the diversity increases dramatically and even includes some sensory modes that we humans lack completely. There is nothing at all
surprising about this, unless you believe that we humans are the measure of all things, but let's not get into that. One sense which we lack
completely is the ability to detect the earth's magnetic field. Some birds
can sense the relative strength and inclination of the gravitational fields
and utilize this information as an aid in navigation. Some aquatic and
mud-dwelling bacteria also have this sense and use it to detect "up"
from "down" by the inclination of the magnetic fields (which continue
at an angle into the earth). At their small size, and buoyed up by the
surrounding medium, gravity is too weak a force to be detected by these
bacteria, but the magnetic fields are detectable due to linear arrangements of small magnetite "inclusions" within the bacteria. Such bacteria
typically use this sense to move "down" into the nutrient-rich sediments.
The birds that detect magnetic fields have been found to have similar
magnetite structures associated with their nervous systems.

Sharks can detect the weak electrical fields generated by the cells of
their prey (most cells have electrically polarized membranes, especially
nerves and muscles). Sharks possess special pore-like organs around
their snouts called the ampullae of Lorenzini, which at close range en-
able them to detect these electrical fields and so allow the shark to
gauge the exact position of their prey in dark or murky water. Electric eels both generate and sense more powerful electrical signals for
purposes of communication, defense, and prey capture.

Animals typically have obvious sense organs connected up to complex nervous systems, which in turn process the sensory information

in various ways. In contrast, other organisms generally lack nerves and highly structured sense organs, but this does not mean that they lack sensation. Remember again that all organisms have the property of irritability (Chapter 1), the ability to respond to stimuli. Since the stimuli must be sensed before they can be acted on, all organisms can sense at least some important properties of their environments.

Prokaryotes, and especially those which have motility, can sense and respond to certain chemicals and chemical gradients in their environment by way of various chemoreceptor molecules located in the outer cell membrane or cell wall of the bacteria. Chemoreception in a liquid medium is essentially the same thing as taste, while chemoreception in a gaseous medium is what we refer to as smell. Chemoreception is believed to be the oldest form of sensation in living organisms. Logically this must be true since the first organisms were either extremely simple cells or precellular aggregations of biomolecules which "lived" by sensing and utilizing some of the chemicals in their environment, while avoiding others (at least at certain concentrations). Such sensing was undoubtedly chemosensory. It would take much longer for the other more complex senses to evolve.

Temperature and light can also be sensed by some motile bacteria and again can lead to attraction or repulsion depending on the species of bacteria. Eukaryotic protists, most of which are motile, also sense and respond to a variety of chemicals, temperature, light, vibration, etc. depending on the species and its requirements. *Euglena*, a photosynthetic protist senses light with its "eyespot" and moves toward lighted areas where it can photosynthesize. *Amoeba* can apparently sense light because it tends to avoid lighted areas. *Stentor*, a semisessile ciliate, can greatly elongate its "body" to reach out into the open water during feeding. It can sense and respond to sudden vibrations in its environment by contracting back into a short rounded shape with less exposure to potential danger. *Paramecium* is said to be negatively thigmotactic, meaning that it responds negatively to touch. When an individual swimming *Paramecium* runs headlong into an object, it immediately backs up, changes its orientation, and swims on. *Paramecium* also engages in a strange form of sex called conjugation which requires that two individuals sense one another (undoubtedly through chemosensation and possibly touch) as appropriate conjugation partners.

Fungi use chemosensation for most of their sensory needs. They recognize their food by chemosensation. Parasitic fungi recognize their hosts through chemosensation. Other symbiotic species such as the mutualistic mycorrhizal fungi recognize the roots of their particular tree

Figure 11.1 The predatory fungus *Arthrobotrys*. Its loop traps quickly swell when a small nematode moves into the loop. Fine extensions of the fungus will then grow and invade the nematode to begin the process of digestion and absorption. (*Figure by Jeff Dixon*)

symbionts through chemosensation. Multicellular fungi typically engage in sex by fusing cells or nuclei from two distinct "mating types" of hyphae, and the recognition of which hyphae constitute the correct mating type is of course done by chemosensation. When the time is right, clustered hyphae of mushroom-producing fungi, as they break through the earth or their host substrate, sense their exposure to light as a signal that they can now proceed to form a mushroom.

Some fungi are predators that catch and then digest soil nematode worms. Among these are species that make special loop-shaped traps along their hyphae (Figure 11.1). These traps are produced in greater numbers if nematodes are sensed chemically in the immediate environment. After the loop traps are produced, should a nematode attempt to crawl through one of the loops, the loop quickly tightens around the nematode holding it in place (due to rapid osmotic changes in the cells of the loop). Subsequently small hyphae grow into the interior of the trapped worm to digest and absorb it. The triggering of the loop trap seems to work through a sense of touch along its inner surfaces. Interestingly, it is believed that female nematodes are more often trapped than males, because the fungal loop may simulate the permanently hooked or curled posterior end typical of most male nematodes. The females may "think" they are crawling into the curl of a male for mating purposes, only to end up a meal for an amazing fungus.

Plants sense and respond to light most often by growing toward light sources. A houseplant placed near a window will (if not regularly turned) end up with most of its leaves tilted toward the window and away from the darker room. The roots and stems of plants typically sense and respond to gravity, the stems by growing away from the source of gravity, and the roots by growing toward the source of gravity. Some tropical vines have the ability to grow toward a nearby dark outline, which often turns out to be a tree that could support the growing vine. In a broad sense of the term, the vine can "see" nearby objects and grow toward them. After making contact with the tree (sensed by touch), the vine switches from growth toward a dark area to growth toward the light from above. As it grows upward, it may spiral around the stem or trunk of the supporting tree, again using its sense of touch.

Some plants can sense the changing 24-hour photoperiods, or light/dark cycles, as the seasons change throughout the year. Some flowering plants, referred to as long-day plants, will only flower when the hours of light per 24-hour period exceed some innate threshold. Long-day plants typically flower in the spring and summer. Others, known as short-day plants, will only flower when the hours of light fall below an innate threshold. These plants typically flower in the late summer and fall, as days become shorter. Shorter days are also the stimulus for the leaves of deciduous trees to stop producing their green chlorophyll. This allows accessory pigments to show through as yellows, oranges, and reds. The short days also trigger the production of enzymes which eventually weaken each leaf's connection to the branch, resulting in leaf fall.

The so-called "sensitive plant" (*Mimosa pudica*) is sensitive to touch. When its leaves are touched with sufficient force, its several leaflets rapidly fold together. Often, the petiole too will droop down when the signal arrives at its base (Figure 11.2). The signal is of an electrical nature, passed from cell to cell, similar in some respects to nerve transmission in animals. The resulting movements of the leaflets and petiole are not due to muscles, but to rapid water loss in special basal cells (in response to the electrical signal). When these special cells loose water pressure, the leaflets fold and the petiole droops. Speculations as to the function of these adaptations include defense from herbivory and reduced exposed surface area during heavy rains and wind, in either case serving a protective function.

The capturing mechanism of the Venus flytrap works initially through a touch sense. Each side of the valved trap has three fine projecting hairs. If two or more of the hairs are bent within seconds of one another, the trap closes. Touching or bending a single trigger hair will not do the job. Any insect worth trapping would be large enough or active enough within the trap to touch and bend two or more of these hairs. The

Figure 11.2 The so-called sensitive plant *Mimosa pudica* before (left) and after (right) having its leaves touched. The leaf folding and drooping occurs within 2–3 seconds. (*Photo by Leon Jernigan*)

bending of the hairs causes certain basal cells to depolarize, similar to nerve depolarization in animals. This signal is rapidly transmitted to large cells at the base of the trap that quickly "collapse" due to a rapid loss of fluid. Since these inflated cells were holding the trap open against the tension of other cell layers, the trap closes as these cells loose their pressure. After initial closure, the inner layers of the trap then "taste" the trapped morsel. If it is food rather than debris, the trap closes even tighter and slowly begins secreting enzymes that will digest the "meal" over a period of a few days. Obviously some fairly sophisticated sensory processes are going on in this "clever" plant.

Throughout this chapter we have seen some of the diversity of sensory abilities both within and across groups, but if you were paying attention, you could not but help notice a fair amount of overlap or unity of sensory modes and responses across those groups. Chemoreception occurs in all of the five kingdoms of life (probably in every species), as does light detection in some forms. Touch sensation, at least in a broad sense, is also present in at least some members of the five traditional kingdoms. Hearing is more restricted, but vibration sensing (a mechanoreception sense not too different from hearing) is more widespread, as in the *Stentor* example earlier. There are, after all, only so many channels of information available to life forms on the planet, and it is doubtful (though possible) that scientists have overlooked any of the major potential sources of sensory information in our complex world.

12

Reproductive and Sexual Diversity

. . . . in no other aspect of their lives do animals show so much variety as they do in their sexual activities.

—Tim Halliday, *Sexual Strategy*, 1980

As stated in Chapter 1, reproduction is one of the fundamental characteristics of all life-forms. Though they all reproduce, the details of how the earth's myriad life-forms accomplish reproduction are likewise quite diverse. Of course not every species is completely unique in the major aspects of its reproduction, but the process does vary widely across the spectrum of earthly life, and can sometimes vary greatly even between closely related species. Certainly to a reproductive behaviorist or physiologist, reproductive diversity would be an obvious parameter of the phenomena of biodiversity.

We will first review some variations that many readers will already be aware of. There are many species of bacteria, protists, and some fungi whose predominant method of reproduction is asexual. Organisms made up of a single cell like most bacteria and protists often simply divide by fission or mitosis into two or more genetically identical, but physically smaller, cells (individuals). All the cells resulting from several such asexual reproductive events, from one original cell, constitute a clone of individuals.

Multicellular organisms like some fungi, plants, and animals can also undergo asexual reproduction, resulting in numerous clones of the original organism, but there is much variation in the way they do it. Sometimes, as in flatworms, the whole parent organism splits or fragments into two or more "chunks" which each grows back into a complete organism. In other groups multicellular buds are produced and released into the environment where they have the potential to develop into new individuals. The *gemmae* of some liverworts, the *statoblasts* of some bryozoans, and the *gemmules* of some sponges are examples of such structures.

Some plants can reproduce new "units" by sending out horizontal above-ground runners (stolons) or below-ground stems (rhizomes). At some characteristic distance from the parent plant, the stolon or rhizome sprouts new above-ground stems and leaves and below-ground root systems. Some plants can reproduce numerous "units" through this type of asexual process. Many species of grasses use stolons to spread across the ground. In landscaping, such grasses are often desirable because a small plug of the grass will spread to eventually cover several square feet, making for more cost-effective landscaping.

Even some trees use rhizomes to spread across the landscape. In the western United States and Canada, aspen trees use this method. Above ground we see stands of what look like separate aspen trees, but below ground the trees are interconnected by a massive system of rhizomes which grew to give rise to all but the original parent tree (which started from a sexually produced seed). The "individuals" of such a cloned stand are all genetically identical. In fact, regardless of the above-ground appearance of many separate trees, the continuous mass of rhizomes below ground implies that such a stand is in fact a single large plant. Some aspen clones cover nearly 100 acres and are believed to be at least 100,000 years old—making them possibly the largest and oldest living individuals on earth, though each above-ground "tree" typically lasts at best a couple of hundred years. There are yet a few other asexual methods found in certain plants, details of which can be checked out in any good general botany text.

Some animals reproduce at least in part through a strange process called *parthenogenesis*. Parthenogenesis is simply the development of an unfertilized egg. The key word here is *unfertilized*. We normally think of only fertilized eggs developing into young, but there are animal species in several phyla that utilize parthenogenesis, and there are several types of parthenogenesis. In the most common type, the eggs of a diploid female are produced by mitosis and so are already diploid (2N). These eggs develop into clones of the mother. Rotifers and some worms and insects use this method most of the time. It obviously requires no mate, and even seems to be more a form of asexual reproduction (the young being genetic clones of the parent). However, because these "eggs" are formed within ovaries in the female reproductive system of the animal, this is often considered a strange form of sexual reproduction. It is an arguable point.

Another form of parthenogenesis which is relatively common involves the development of haploid (N) eggs, which were the product of meiosis. Most commonly, these unfertilized haploid eggs develop into males, which themselves are haploid. This is known to occur in both rotifers and some insects. All drone bees are produced by this method. Because these males are totally haploid, they produce their sperm by a mitosis-like division, rather than by meiosis. This means that all their sperm are identical. This is totally at odds with what most students learn about meiosis and sperm production in general biology classes. Perhaps introducing this variation to introductory students would just be too confusing. There are yet more variations on parthenogenesis, but they are very rare indeed, so I will leave you to wonder what they might be and move on.

In the realm of "true" sexual reproduction, the diversity is even greater. Sex is usually associated with reproduction in the minds of most people (even biologists), but the two phenomena are not the same thing and are not universally connected. *Paramecium* and many other ciliates engage in a sexual process called conjugation. Two paramecia, neither of which can be said to be male or female, come together and fuse their cell membranes creating a cytoplasmic bridge (Figure 12.1). Then meiotic nuclear divisions occur resulting in several haploid nuclei in each individual. One of the haploid nuclei from each individual moves through the cytoplasmic bridge connecting the two organisms and fuses with one of the nonmigrating nuclei in the recipient individual to create a diploid nucleus. After this is accomplished, the two individuals break their cytoplasmic connection, and move apart to continue their activities. Later on they will divide by fission, just as other Paramecia do which have not engaged in conjugation.

Conjugation involves genetic reshuffling and recombination. Each *Paramecium* goes away from conjugation with a new genetic makeup, half of which came from the other partner, but neither is what could be termed "pregnant." Though they will later split by fission to create two new individuals, paramecia that have not conjugated recently can do the same thing. Here sex and reproduction seem largely unconnected, fission results in reproduction while conjugation results in new genetic combinations, but no obvious connection between the two processes is visibly apparent.

Again, paramecia don't have obvious sexes, so in this respect they have less biodiversity than species that do contain distinct male and female forms. Some of the multicellular algae produce haploid gametes, but they still don't express what could be called maleness and femaleness.

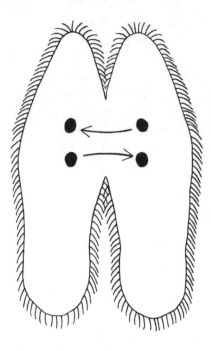

Figure 12.1 Two paramecia in the process of conjugation. The two individuals have aligned and formed a cytoplasmic bridge. In each individual a diploid nucleus has undergone several divisions to form haploid nuclei (most not shown). The last two haploid nuclei to form (shown in black) will engage in a genetic exchange where one haploid nucleus from each individual will migrate into the mating partner to fuse with the nuclei that did not migrate (fertilization). After this fusion, the new single nuclei in each individual will be diploid. The partners will then separate and go their separate ways. (*Figure by author*)

The gametes look identical and can't be distinguished as sperm or eggs. There is a genetic factor present which makes the gametes from some individuals "positive," and those from other individuals "negative." Positive gametes can only fertilize negative gametes. Here we have the production of sex cells (gametes), but still no males or females. Some of the fungi use a similar system. Other algae and fungi have evolved more distinct mating types that can be labeled as the male and female sexes.

Back "down" in the prokaryotes, the situation is quite different. Though prokaryotes typically divide asexually, they do have a strange form of sex also known as conjugation. Bacterial DNA is mostly in the form of one long circular "chromosome," but some is contained in smaller circular units called plasmids. Bacteria containing certain plasmids can form cytoplasmic bridges (pili) to other bacteria lacking those plasmids. The plasmid carrier replicates its plasmid and "injects" a copy (through the hollow pili) into the bacteria lacking that plasmid.

Sometimes plasmid DNA can become incorporated into the larger chromosome, and cells of this type can replicate all or part of their chromosome and inject the copy (or part of the copy) into cells which lack such plasmid-derived DNA in their chromosomes. In all such exchanges, a cell with more DNA is replicating part of its DNA and injecting it into another bacteria which has less DNA. There is no true haploidy or diploidy as in eukaryotes, only groups of genes moving from one cell to another. The process reminds one almost of infection of one cell with the DNA of another, in a manner reminiscent of how viruses inject their DNA into a host cell. It also seems reminiscent of the old adage about "not putting all your eggs in one basket." The injecting cell doesn't lose any of its original DNA in this sexual process, but it does disperse copies of some of its DNA into other bacteria, where they may live on should the "parent" cell meet disaster. Most surprisingly, prokaryotes can do this across species lines, even when phylogenetically distant from one another. This is almost certainly yet another face of genetic or biological selfishness.

Most complex animals and plants have clear sexual differences at many levels. A male is defined as the sex which produces sperm, while a female is defined as the sex which produces eggs. Sperm differ from eggs in three universal ways: (1) sperm are smaller than eggs, (2) sperm can move under their own power while eggs cannot, and (3) sperm are produced in far greater numbers than eggs within a species. Remember that some protists, algae, and fungi don't produce eggs or sperm, only morphologically identical haploid nuclei or gametes that can potentially fertilize one another. Sex certainly started out this way (after the simpler bacterial gene injections mentioned earlier) followed later by the evolution of two distinct sexes and a more direct connection between sex and reproduction.

We already know the basics of sexual reproduction in our own species. We are a dioecious species, meaning that we have separate sexes—male and female. We mate, and the female houses the developing young until birth—typical, or so we think, for animals in general. One animal that varies in a dramatic way from this pattern is the seahorse. Seahorses are actually fish, bony fish to be more precise. In seahorses, the mating process involves a delivery of the female's eggs into a brood pouch on the male's belly. After the eggs have been passed to the male, he releases sperm into the brood pouch and fertilization occurs there. The fertilized eggs develop within this male pouch until they are baby seahorses, after which they are expelled by rhythmic contractions of the male's pouch. In essence, it is the male that gets pregnant and later gives birth to the young!

In humans and seahorses, eggs are fertilized inside the body of one of the parents, an event termed "internal fertilization." Of course, many species such as some fish, algae, jellyfish, polychaete worms, etc. use external fertilization in which eggs and sperm are released into a watery environment where they will hopefully meet and fuse. In internal fertilization, sperm transfer (typically into the body of the female) can be either direct or indirect. Direct sperm transfer is the method used by humans and any other animal species in which the sperm tubes exit the body through the penis, used in copulation.

Some animals use another method termed "indirect sperm transfer." In this method sperm exits the body of the male first, and is transferred into the female a bit later. In spiders, the male deposits his sperm on a small horizontal "sperm web" which he spins just for this purpose. Then he turns around and sucks the sperm into two modified appendages near his mouth called *pedipalps*. He is then ready to deliver his sperm to a female from these filled pedipalps. In salamanders, the male deposits his sperm in a club-shaped packet called a *spermatophore* which he sticks to the ground. He then leads his chosen mate over his spermatophore; if cooperative, she will suck the spermatophore into her genital slit where the sperm will escape to fertilize her eggs. There are many variations of indirect sperm transfer in the animal kingdom, and though less common than direct sperm transfer, it is not an uncommon practice.

Many animals and plants are monoecious, meaning that each individual houses both a male system for making sperm, and a female system for making eggs and possibly for housing developing young. Earthworms and snails have this arrangement. Though sperm and eggs are present in the same body, most monoecious animals avoid fertilizing their own eggs with their own sperm. This would be the ultimate in "inbreeding," which most of us know is not typically a good way to insure the health and viability of offspring. Instead, two monoecious individuals come together and mate. When two earthworms or snails mate, each is delivering sperm to the other, and each is getting its eggs fertilized by the sperm of the other individual. A big advantage of this arrangement appears to be that any other member of your species is a potential mate. This could be an important advantage if your species is sparsely spread over the landscape such that meeting another of your species is a rare event.

There are other animals which can be both male and female, therefore monoecious, but not at the same time. Several species of marine fish and invertebrates are *sequential hermaphrodites* (hermaphroditic is another term for monoecious). Some are *protandrous*, starting life as a male and later changing to female, while some are *protogynous*, starting life as a female and changing later into a male. In these animals, not only does internal function and physiology change from sperm production to egg production (or the reverse), the behavior and outer morphology typically also changes to signal its new sexual identity (Figure 12.2).

In some species of polychaete worms (phylum Annelida), the typical worm (an atoke) eventually buds asexually from its posterior end a new worm (the epitoke), which though genetically identical to the anterior parent, is morphologically distinct and has functional sex organs

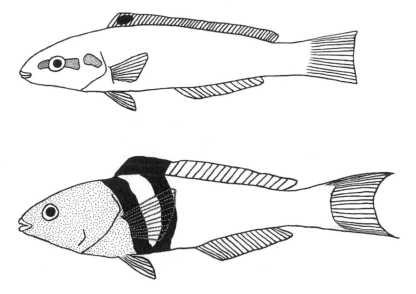

Figure 12.2 Sex change in the Blueheaded Wrasse (*Thalassoma bifasciatum*). If a male should die or disappear from an area in the reef, the largest female in the area (above), can transform in days into what is called a termal male (below). The female is mostly yellow, while the male will have a blue head and a blue-green to yellow-green posterior, with a stripe of white between the two black vertical stripes. Of course the behavior will change along with the sex of the transforming female. Such sex change from female to male is called protogyny. Some other fish species change from male to female with age, a process called protandry. (*Figure by author*)

for forming sperm or eggs. Eventually this sexual worm breaks free of the original atoke and swims off to locate a mate or simply rupture in a coordinated manner along with many other sexual worms, releasing gametes into the seawater where external fertilization will occur. The atoke never develops sex organs and so never engages in sexual activity, only the budded epitoke individual is capable of sexual reproduction.

Another strange twist occurs in monogenetic fish flukes of the genus *Gyrodactylus*. These flukes are monoecious. After mating, sexually produced embryos develop in the uterus of each individual and eventually are born alive (viviparity). This is not so unusual, even for simple worm parasites. What is unusual is that just before birth, the fetal fluke is already "pregnant" with another developing embryonic fluke, which has inside it a smaller developing embryonic fluke, which has inside it yet another even smaller developing fluke. The flukes are "nested" inside each

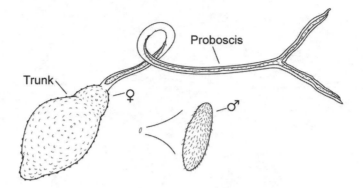

Figure 12.3 The benthic marine spoonworm *Bonellia.*
Note the drastic size difference between the female and
the tiny dwarf male (enlarged for clarity) that comes to
live within her reproductive tract. (*Figure by Jeff Dixon*)

other in a manner reminiscent of the popular "stacking dolls" of Europe
and Russia (Matryoshkas). The count is four juvenile flukes nested inside
the "mother" fluke. Only the first to mature is the direct result of sperm
and egg fusion (fertilization). The latter three nested juveniles are actually asexual clones of the first. In sequence each fluke develops rapidly
and gives birth to the clone within its body, except for the last which
contains no remaining clones. Then each fluke becomes sexual mature
and eventually mates with another fluke. This strange phenomenon is
termed *sequential polyembryony.* Polyembryony (many embryos resulting
from one zygote or fertilized egg) though rare, is known from other
animal groups such as some of the parasitoid wasps, but the nested sequential polyembryony that occurs in *Gyrodactylus* is a real rarity. I think
you will agree it also qualifies as a real "believe it or not" oddity of
nature.

A few animal species have evolved another strange twist that has been
termed "sexual parasitism." In most cases of sexual parasitism the male is
much smaller than the female, and is often referred to as a dwarf male.
Once a dwarf male locates a female he either attaches permanently
to the female as in some deep-sea anglerfish, or enters some chamber
of the female reproductive tract as in the spoonworm *Bonellia,* living
there ready to release sperm when needed by the female (Figure 12.3).
In either situation, the male becomes a physiological parasite of the
female, deriving all his nourishment from her blood or tissues. Having
her eggs fertilized is apparently more than a fair exchange from the

standpoint of the female, so in net balance, the males are not parasites in the sense that a leech or a flea is a parasite.

A final topic worthy of mention is sex determination, that is, in dioecious species, what determines whether a developing individual will be a male or a female? Most of us know that in humans and most mammals there are two different sex chromosomes, the X and the Y. If a zygote receives an X chromosome from both parents, the embryo will develop into a female. If the zygote received a Y chromosome from the sperm (it couldn't get a Y from the mother), then it will develop into a male. Though some other animals use this system, there are many other variations on sex determination.

We have already seen that in bees, wasps, and some rotifers sex is determined by whether the developing egg is diploid or haploid, with only the haploid eggs developing into males. The diploid eggs, regardless of whether resulting from maternal mitosis or from fertilization by a male's sperm, will develop into the females of the species.

In some fish, reptiles and insects, and in all birds the situation is reversed and it is the male who has two identical sex chromosomes. In birds the males have two Z chromosomes (ZZ), while the females have the WZ condition. In some lizards and turtles, the sex of the developing embryo is determined by the temperature at which the eggs develop, with cooler temperatures favoring a female outcome and warmer temperatures favoring the male. Here the sex is not genetically determined, but rather environmentally determined.

An even stranger case of environmentally determined sex is found in the marine spoonworms of the genus *Bonellia*, one of those groups with dwarf males mentioned earlier. Sex determination is really unique in these worms. Their life cycle involves the development of a ciliated larval stage. If the larvae settle on the bottom away from other members of the species, they develop into females. If, however, they settle on, or soon contact, the extended prostomium of a female (Figure 12.3), they usually develop into dwarf males and become sexual parasites within the female. Obviously some chemical signal is received from the female body that directs the embryo into the developmental path leading to maleness.

Though I have tried to minimize this, there is a certain amount of unavoidable subject overlap in different chapters of this book, and you will find even more information pertaining to the biodiversity of sex in the next chapter which deals with the diversity of life cycles (just as some life cycle details appeared in this chapter). We have only covered here a few of the many interesting variations in reproduction and sex. There are other reproductive diversity details covered by the terms oviparous, ovoviviparous, polygamy, polyandry, hypodermic impregnation, sexual selection, genitalia morphology, etc. I trust we have covered enough material here to make the case that within the area of reproduction alone, there exists a plethora of biodiversity.

13

Life Cycle Diversity

Perhaps no single group of related organisms on earth possesses greater
life history variability than do members of the Nematoda.
                    —A. O. Bush et al., *Parasitism: The Diversity and Ecology of*
                                                        Animal Parasites, 2001

As mentioned elsewhere in the book, some organisms do not look the
same throughout an entire life cycle. Bacteria and many protists reproduce exclusively by asexual cell division and do look and behave basically
the same in any stage of their individual growth (unless they regularly
form dormant cysts in their life cycles). Though we may think that human babies look fairly different from adults, and although it was argued
in Chapter 9 that the changes in morphology with age are a valid aspect
of biodiversity, babies are certainly "humanoid" with all the same parts
that adults have. Human babies could not be mistaken for a member of
any other species, not even a closely related species such as the chimpanzee. We do indeed go through some developmental stages in the
uterus which start out very simple and unhuman-like until we take on
an unmistakably human form by about 7 weeks of development.

A great many species do not go through a generation so simply. The
organism that causes malaria (*Plasmodium falciparum*), though it is a
single-celled protist, goes through one of the most varied and complex
of all life cycles. It parasitizes both humans and mosquitoes, and it
progresses through several distinctive forms or morphologies in each of
its two hosts. It must pass into the mosquito to enter another human,
and it must be passed to a human before it can enter another mosquito.

As you can see in Figure 13.1, the parasite enters the human as an
elongated sporozoite form which quickly enters a liver cell where it grows
and divides into numerous tiny cells called merozoites. These merozoites

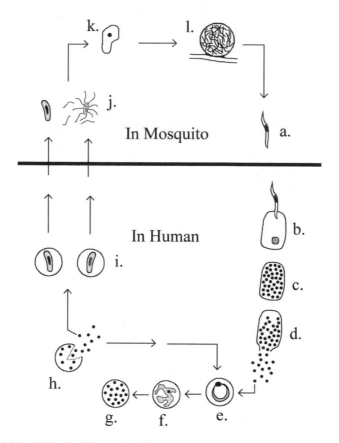

Figure 13.1 Malaria life cycle (*Plasmodium falciparum*). (a) sporozoite in mosquito salivary glands, (b) sporozoite locates a liver cell and penetrates, (c) *Plasmodium* cell has grown and undergone multiple fission—the liver schizont stage, (d) liver cell ruptures spilling merozoites into circulation, (e) after entering a red blood cell, the merozoite grows into a ring-form trophozoite, (f) cell grows into an amoeboid trophozoite, (g) trophozoite undergoes multiple fission into merozoites—the red blood cell schizont stage, (h) the red blood cell ruptures spilling merozoites into the blood plasma—these enter more red blood cells, (i) some of the merozoites on entering new red blood cells develop into male and female gametocytes, (j) some gametocytes are sucked up with the blood meal of a new mosquito where the male gametocyte divides to form several tiny male gametes, one of which will fertilize the female gametocyte, (k) the ookinite develops from this fertilization and burrows through the gut wall of the mosquito, and (l) the ookinite develops into an oocyst within which multiple sporozoites form, escape, and move up to the salivary glands. (*Figure by author*)

escape from the now dying liver cell to enter circulation and invade red blood cells (RBCs). Here they undergo a pattern of growth, development, and asexual reproduction in which several sequential forms of the parasite are recognizable. First develops the ring stage, followed by the amoeboid stage, followed eventually by the stage in which a multiple asexual division occurs (the schizont stage). The RBC finally ruptures to spill out the numerous offspring of this reproductive event, also known as merozoites.

These RBC-produced merozoites enter other RBCs to either repeat the process, or to grow into one of two sexual stages: a female macrogametocyte, or a male microgametocyte. If RBCs containing macrogametocytes and microgametocytes are sucked from the body into the stomach of a mosquito, the microgametocyte divides into several very tiny and elongate microgametes. Microgametes attempt to fuse with or fertilize any available macrogamete (now freed from its surrounding RBC). The fertilized cell which results then takes on an amoeboid form (the ookinite) and burrows through the wall of the mosquito stomach where it forms a cyst-like structure termed an oocyst. Many sporozoites then form within each oocyst. These sporozoites break free and some move up to the salivary glands of the mosquito from where they can move out of the mosquito and into the next human victim the mosquito bites.

This is a truly amazing life cycle with great variety of forms and processes occurring in succession. Some forms live in the human while some live in the mosquito, two very different environments. In humans the parasites live within cells, while in the mosquito they don't. Some forms can move under their own power (the microgametes and ookinite) while most cannot. Some forms carry out the sexual steps in the life cycle, while other forms carry out asexual reproduction. There are sensory and recognition differences in the various stages because injected sporozoites sense, recognize, and enter only liver cells, merozoites recognize and enter only RBCs, the microgamete recognizes and fertilizes the macrogamete, and the ookinite recognizes and burrows through the mosquito gut wall. Finally, there are undoubtedly complex metabolic differences between the various stages which grow in different environments (liver cells, RBCs, mosquito stomachs, etc.). It is frankly hard to imagine how a life cycle could be more complicated.

Across the whole spectrum of life there exists an amazing range of diversity both within life cycles (as in the Plasmodium example earlier), and between the life cycles of various groups. We could not begin to look at more than a small fraction of life cycle diversity in a single chapter (there is also some pertinent material in Chapter 9), but to support the point further, we will look into a few more examples.

Members of one group of the Protoctista, the cellular slime molds, have relatively unique life cycles in which the active feeding stages are separate unicellular amoeboid cells which creep about the environment (moist soils, rotting vegetation, etc.) feeding on bacteria. When the food supply dwindles or other adverse conditions threaten, the many separate feeding amoeba in an area will aggregate to form a migrating stage which looks something like a slug (the pseudoplasmodium stage). This stage moves in a manner that at least suggests a single multicellular organism leaving the area for "better pastures." If the environment and/or the slug dries out, the mass of cells stops and reforms as a slim vertical stalk with a swelling at the top. Cells within the upper swelling (the sorus) form spores which are released to scatter over the new area where they eventually "hatch" to release new unicellular amoeba which again begin the search for food in their new surroundings.

This life cycle varies in numerous ways from that of *Plasmodium*. The slime mold cycle has fewer stages, is not parasitic in any of its stages, includes a multicellular stage unlike *Plasmodium*, and does not include a sexual fertilization event as did *Plasmodium*. Obviously, it also varies greatly from the human life cycle. It also varies among its stages in that the motile amoeba are feeding unicells, the motile multicellular slug form is nonfeeding, and the multicellular stalked form is sessile and nonfeeding.

Let's next look at one of the common marine green algae known as *Ulva*, or sea lettuce (Figure 13.2). The large flattened leaf-like bodies of *Ulva* are attached to the substrate in shallow marine habitats. Strangely there are two fundamentally different forms of *Ulva* present which cannot be distinguished visually. One form is called the gametophyte whose cells are haploid. It is called the gametophyte because it can form gametes (sex cells) which escape and fertilize one another in the seawater. The gametes are also haploid as you would expect, but were formed by mitosis (unlike the situation in most animals).

After fertilization, the now diploid zygote develops into another *Ulva* organism which looks just like the gametophyte, but whose cells are diploid. This form is called the sporophyte because later on it will be capable of forming spores. The spores are formed through meiosis and come out haploid. These spores then develop back into another *Ulva* gametophyte. Both the gametes and spores are motile haploid cells, but again, the gametes form through mitosis, while the spores form through meiosis. The gametes of *Ulva* all look alike, so there is no male or female

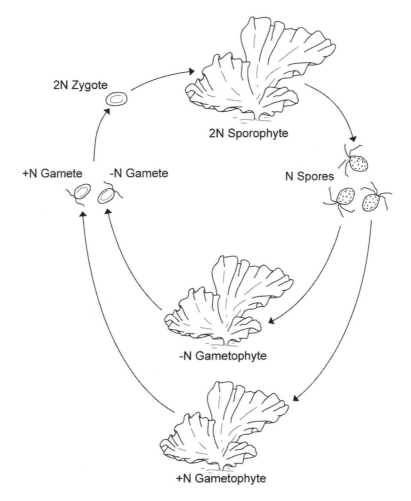

Figure 13.2 The life cycle of *Ulva*, a type of marine green algae. Further explained in text. (*Figure by Jeff Dixon*)

gamete as in most sexually reproducing organisms, including even *Plasmodium*. This life cycle is quite distinct from our earlier examples. There are other algae (and plants) in which the sporophyte and gametophyte forms are themselves quite distinct in morphology, unlike the situation in *Ulva*.

Because parasites are so "neat" and numerous, let's include one more parasitic example (after all—most of the world's species are parasites) from the animal kingdom, the sheep liver fluke *Fasciola hepatica* (Figure 13.3). Flukes are flatworms of the phylum Platyhelminthes, which also

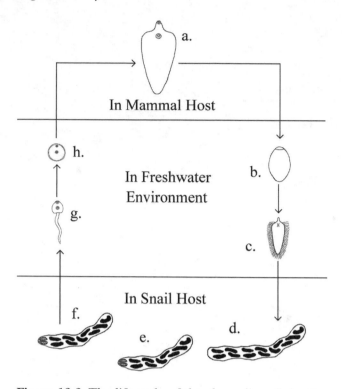

Figure 13.3 The life cycle of the sheep liver fluke *Fasciola hepatica.* (a) The adult fluke, (b) egg, (c) miricidium, (d) sporocyst, (e) mother redia, (f) daughter redia, (g) cercaria, and (h) metacercaria. Explained further in text. (*Figure by author*)

includes the free-living Planaria flatworms that most biology students encounter somewhere in their studies. The adult flukes live in the bile ducts (in the liver) of many of the larger hervibores such as sheep, cattle, goats, and deer, and sometimes they find their way into human hosts as well.

The adults mate and produce eggs that make their way into the bile, which mixes with the material in the intestines, and is finally passed out with the feces. If fecal material containing fluke eggs ends up in bodies of freshwater, ciliated larval forms known as miricidia hatch from the eggs and seek out freshwater snails. If such snails are located, the miricidia burrow into their bodies. Inside the snail the larvae grows and develops into a sporocyst stage. The sporocyst grows and subdivides its internal tissue into many "mother redia" which break out of the sporocyst and continue to grow and subdivide their internal tissues into several daughter redia stages, which in turn break out of the mother sporocysts. The daughter redia undergo yet another round of growth and internal subdivision into several cercaria which break out not only from the redia, but also from the snail's body. The cercaria swim away and soon form a metacercarial cyst stage in the water or on aquatic vegetation. If a large mammal drinks the water containing the cysts or ingests cysts along with aquatic vegetation, the cyst will hatch in the mammal's gut and a larval fluke will move into the animal's bile ducts to grow into another adult fluke, yet another amazingly complex life cycle in terms of morphology.

This cycle is also complex in other ways. Sporocysts, redia, and metacercaria stages are essentially nonmotile. The adult fluke can move about in the bile ducts using its suckers and muscular contractions of its body, the miricidia swim through the water with the aid of their cilia, and the cercaria stage swims with its tadpole-like tail. There is sexual reproduction in the mammal and asexual reproduction in the snail. The cercaria and miricidia, though short-lived, are free-living, while the other stages are all parasitic.

Being out in the environment, the miricidia and cercaria are aerobic in their metabolism, while the adult flukes in the bile ducts carry out only anaerobic metabolism.

There are so many life cycles from which to choose for illustrating life-cycle diversity. We could have mentioned the familiar butterfly/caterpillar cycle, or the tadpole/frog example, or others which most people would be familiar with. Humans were mentioned at the start of this chapter, and our simple (by comparison) life cycle is but one of many which lie along a spectrum of great biodiversity. From the relatively simple nonsexual fissions of most bacteria, to the unbelievable complexity of the *Plasmodium* life cycle, biodiversity abounds in the details of such various life cycles.

14

Behavioral and Cultural Diversity

Different species display a vast array of behaviors; and within a species, males and females often behave quite differently.
—J. W. Grier & T. Burk, *Biology of Animal Behavior*, 2nd ed., 1992

Organisms are not just material structures with distinct morphologies, they are also what they do. Organisms are alive because of their ongoing processes and activities. Though some organisms such as most plants and fungi don't appear to "do" much, others such as most protists and animals regularly engage in complex and varied activity that we refer to as behavior. It has been suggested that when you view a lion in a zoo, especially one that has lived its entire life in captivity, you are not truly seeing a lion, only the empty shell of a lion. Lions in captivity, especially years ago when they were kept isolated in small cages, do not carry out typical lion behavior. They have had no opportunity to learn the intricacies of normal lion behavior from a normal functioning social unit (a pride). We will come back to learned behavior later. Most behavior is actually innate, meaning that the genetic makeup of the organism largely determines what kinds of behavior that organism will be capable of, and how those behaviors will be "structured."

Genes are often analogized as the blueprints for an organism's physical structure, but many are also the blueprints for what the organism will do in terms of behavior. Genes of course don't directly control behavior, but they do determine much of the basic wiring layout of the nervous system, which itself (along with other gene products like muscles, support systems, and glands) is the direct mediator of behavior in most animals. Less is known about how protists control and regulate their behavior, considering that they lack nerves and are themselves but a single cell. It is nevertheless assumed that by some pathway, genes do fashion and regulate their behavior, and some variations in protist and animal behavior are known to be the effects of single genes with two or more alleles. Of course environment is important, even in innate behavior. Certain environmental conditions and factors are often essential in setting the stage for, and in triggering, the "normal" behaviors of respective species.

Behavior is one of those biological phenomena that can be referred to as an "emergent property." Emergent properties are those properties of a system which are not predictable or additive from knowledge of the respective "parts" and lower levels of organization within the system. In other words, studying the biochemistry and morphology of an animal does not tell you much about what behaviors that animal will carry out when alive and functioning in nature. Sure, the presence of wings in a bird might suggest flight, but that is only because we already know from observing *living* birds that wings serve that function. Some birds use their wings in other amazing ways which are not predictable from watching birds in general.

In the film series *Trials of Life* by David Attenborough, the courtship episode includes a species called the riflebird. In courting the female, the male carries out some singularly unexpected wing movements and postures (as do numerous other species) that are both strange and beautiful to watch. No knowledge of this bird's physical body could have predicted such complex behavior, thus it is an emergent property, springing into existence only in the complete living animal in the proper environmental context. Most behavior is emergent to some extent. No matter how much we know of dinosaur skeletons and muscle attachments, we will never know the details of how dinosaurs could move or how varied their behavior was. The same would be largely true if we had an intact, but dead, dinosaur carcass to study.

Behavior exists and varies on a different and separate scale from either morphological or genetic diversity. For starters, individuals of a single species may carry out numerous distinctive behaviors such as foraging, feeding, defense, courtship, mating, grooming, etc. Also, there can be a high degree of behavioral disparity between species that vary only slightly in their morphology and genes. A perfect example of this phenomenon is found in our immediate phylogenetic group which includes chimpanzees, bonobos, and humans. We now recognize two species of chimpanzee-like primates, the chimpanzee and the bonobo. Behavioral diversity is obviously great between humans and either of these two close relatives, but it is also very great between the chimps and bonobos, which we judge to look quite similar morphologically. All three species vary from each other by less that 2 percent at the genetic level.

While bonobos are most often quite and peaceful, chimps are more often engaged in restrained or all-out displays of aggression, dominance, fear, etc. Bonobos are also known for the frequency and variety of their sexual encounters. There are seemingly few if any rules for sexual engagements and activities among the bonobos. Sex appears for them to be the major means of maintaining peace and social unity. Chimps are downright puritanical by comparison, generally using more restrained physical contacts and aggression to mold their societies. If someone who was not familiar with the slight morphological differences between chimps and bonobos was given a clear description of the typical behaviors of each species and then allowed to observe a social group of one of the two species, it would likely take only an hour or so of observation to determine with great certainty which species was being observed.

There are even cases in which the behavior of two or more species can be different while their morphology is not. Sibling species (mentioned in Chapter 2) exist when two or more species, which are usually closely related, have not diverged noticeably in outward morphology, but often do show some consistent behavioral differences. If the species are sexual, then at least in mate choice their behavior is species-specific, since they would not normally mate with members of another species, no matter how similar their morphology might be. Often there are differences in their calling signals and/or courtship behaviors that keep the sibling species reproductively isolated. Two bird species, the dusky flycatcher and the gray flycatcher, are almost completely identical in their morphology, but can be separated by their songs, which presumably aid the birds in identifying which individuals are potential mates of their own species. Protists are rich in sibling species "complexes." The two ciliate genera *Paramecium* and *Tetrahymena* each contain numerous sibling species which recognize each other chemically and do not hybridize across species lines.

Conversely, two species that share little morphological similarity may both perform behaviors which are very similar in general appearance. A good example of this phenomenon would be the swimming behavior of a sea snake and an eel. The snake is a reptile with terrestrial origins, while the eel is a bony fish with a long ancestry of nothing but swimming ancestors. Nevertheless, sea snakes adapted to ocean living, laterally flattened their bodies slightly, and swim with the same general side to side body undulations that eels use in their swimming.

Now to the biologically challenged individual, sea snakes and eels may not appear that different, but to anyone with a good eye and basic knowledge of the vertebrate classes, there would be several obvious and drastic differences between these two animals such as presence/absence of gills, scale structure, skull design differences, etc. Their similar swimming behavior is yet another example of convergent evolution which was discussed back in Chapter 1, and there are many other behavioral examples of this phenomenon. This example further illustrates that behavior does not vary along the same scale as either morphology or genetics, both of which would differ greatly between sea snakes and eels.

Behavior can even vary dramatically within a single species. The insect castes of social insects illustrate this phenomenon nicely. Soldier termites do not behave like the queen, the king, or the workers of their species. In many species male behavior may vary dramatically from female behavior. The behavior of the young may vary greatly from that of older individuals, especially so when the young are morphologically distinct from the adults as is the case in many invertebrate life cycles (caterpillar/butterfly, etc.). Of course there are countless examples of behavioral differences between males and females of the same species, and hopefully most readers can supply several examples of their own on this point.

Many animals show age-related behavioral diversity, but perhaps the best illustration of this phenomenon can be found in the common European honeybee. Each worker bee only lives for 30–35 days, but during that short lifespan the workers instinctively and sequentially shift from one task to another. At 1–2 days of age the workers clean the hive. From 3 to 11 days of age they feed and care for the larvae and queen. From 12 to 18 days of age they guard the hive, cool the hive (if necessary), and build wax cells (the honeycomb). Finally, from 19 days of age until death, they leave the hive to forage for food and scout for new food sources. In honeybees, you can generally estimate the age of a worker bee simply by observing what it is doing.

Animals can also alter their behavior in response to learning. They can be sensitized to react strongly to stimuli they did not react to previously. They can habituate to repeated stimuli, eventually failing to respond to stimuli they had formerly reacted to. Some animals can learn to associate a stimulus with a reward or punishment and react to that stimulus, whereas before the associative learning (conditioning) took place that same stimulus would have been largely ignored by the animal. As adults, humans often react differently to some forms of art and entertainment than they did as juveniles. We will speak of learning and humans later in the chapter. A general rule is that the more complex the animal, the more unpredictable its behavior, and the more likely that behavior will change in some ways over the lifetime of that individual.

Where behavior is largely learned, behavioral diversity is possible among species, within species, and even within individuals. Animal trainers long ago showed that animals could learn new and novel behaviors that their natural counterparts did not engage in. Later it was recognized that animals in nature can "invent" new behaviors which, if useful, may even be passed down the generations as a true culture of learned behavior. Roger Payne's work with humpback whales has shown their acoustic communications (songs) to be learned. He also found that these songs change on a time scale much shorter than could be explained by any genetic changes. In fact, individual whales alter their song as they mature in ways that suggest individual invention and whimsy.

Chimpanzees have been extensively studied in Africa as to their cultural traits and complexities. Chimps are not our ancestors, rather they are a relatively new species which has evolved almost as rapidly as humans from a common ancestor which existed some 6 million years ago. Chimps are spread across central Africa in several isolated areas. Interestingly, almost every isolated population shows unique culturally based behaviors. In the forests of the Ivory Coast in western Africa, chimps engage in the gathering of edible nuts which they crack open with stones. They place the nuts in shallow depressions on the exposed bracing roots of certain trees and then hammer them with the stones. Chimps in other areas of Africa do not have or use this technology. Chimps in the Gombe area (those studied by Jane Goodall) engage in termite "fishing." They break small openings into termite mounds and then insert slender plant stems into the tunnels of the mound. As they wiggle and slowly twist the stems, worker and soldier termites grab the stem with their mandibles. The chimps then pull out the stems along with several attached termites and promptly eat the termites. Again, this cultural technology is not found in other disjunct chimp populations. Such behaviors seem by all indications to be passed on by observational learning as the young watch older individuals perform these behaviors. In the case of the nut cracking, there may even be some teacher–student learning occurring, with adults purposefully helping the young to acquire the skills. As in human cultures, chimp culture is diverse from one population to another, and there are other examples of this diversity in addition to the two mentioned earlier.

Our own species is the supreme example of how learning contributes to behavioral diversity. Though some, like E. O. Wilson, have pointed out and stressed the underlying similarities of almost all human cultures, there is undeniably an outward expression of diversity in most of the behavioral events among and within human cultures. We perhaps vary

most in those behaviors we perform for amusement. How many games have humans invented? Certainly the number is in the thousands, and though some may share several features, chess is in no way similar to tennis, nor is golf similar to poker. We of course vary in more biologically pertinent areas such as our sexual behavior. How many elements of foreplay and positions for "coupling" have been devised by our species? Even though many of these variations may be little-used by the mass of humanity, there is most definitely a lot of variation in our sexual behaviors both within and among cultures.

Some individual humans have varied tremendously in terms of their behavior over the course of a single life. At least outwardly, some humans seem to have moved from heterosexual to homosexual (and vice versa). Some who cannot swim will learn how to swim very effectively. Some who devote much time and behavior to religious practices turn atheist and cease engaging in these practices (and vice versa). Some who once practiced one profession have changed tracks into totally different and unrelated professions—and so on. You can undoubtedly think of several other examples where a person's behavior patterns changed over their lifetime as the result of their experiences or environment. Only in the most primitive and rigid of human societies are such changes viewed as unusual. Such behavioral shifts are typical of humans in much of the modern world, and they add yet another layer of behavioral diversity to our species.

Across the whole animal kingdom, behavior varies tremendously. Behavior is so varied that there seems to be no single obvious scheme for naming and categorizing the multitude of behaviors found in nature. One possibility is to rank behaviors according to their complexity, or the complexity of nervous system processing required to perform that behavior. Table 14.1 illustrates one traditional listing of these levels. Some behaviors at a lower level of complexity may appear more complex than one at a higher level. For example, the construction of an orb web by a garden spider is a very complex and lengthy process involving at least hundreds of discrete movements of legs and spinnerets (the short appendages which produce the silk), while knowing by associative reasoning that almost any grocery store in the United States stocks peanut butter in close proximity to jams and jellies seems simple by comparison. The distinction is that the nervous circuitry required to learn and associate is almost certainly more complex that the spider's "hard-wired" circuitry dedicated to web construction, a behavior which is innate and far less flexible than most forms of learning and reasoning.

Table 14.1 Complexity Levels (Categories) of Behavior

Kinesis: An undirected locomotor response to a stimulus such as light, sound, moisture, chemicals, etc. For example, woodlice move about more in dry environments than in humid environments.

Taxis: A whole-body orientation response to a directional stimulus. May often be coupled with locomotion, but not always. The response may be positive if the orientation is toward the stimulus source, as when a male moth orients toward the female pheromone source (the female). The response may be negative if orientation is directly away from the stimulus source, as when *young* salmon orient their swimming downstream, away from the source of the current. Finally the response may be at some specific angle to the stimulus source, as when lizards bask at changing angles to the sun's rays depending on how cold or warm they are.

Reflex: An automatic, rapid, simple response; often involving only a part of the body in animals. Examples would include coughing, knee-jerk reflex, eye-blink reflex, salivation by a hungry dog when food is detected, etc.

Innate behavior patterns (or fixed-action patterns): An innate pattern (not learned) consisting of two or more steps (muscular actions) which are performed in an exact sequence. Examples would include the spinning of a web in spiders, flying in butterflies, mating in sharks, construction behavior in paper wasps (of the nest), etc. No learning is needed for these respective animals to carry out these complex patterns of behavior.

Learning: An acquired change in, or addition to, an individual's behavior as the result of experience. The change is the result of some form of memory. There are several subcategories of learning which can also be arranged from simple to complex:
- Habituation
- Sensitization
- Imprinting
- Latent learning
- Conditioning (classical and operant)
- Observational learning (imitation)
- Teacher—student learning

Reasoning:
- Association Reasoning: Recognizing similar yet slightly different situations as having similar characteristics, and behaving appropriately based on learned information. Concept formation.
- Insight Reasoning: Mentally "seeing" solutions to new problems and situations

Behavior can (and has) also been grouped "functionally," that is by the function that it assumes in respective animals. The list here is rather long (Table 14.2), and somewhat unsatisfactory. Some functions are widespread among animals such as feeding behavior and mating

Table 14.2 Functional and "Other" Behavioral Groupings

1. Foraging behavior: hunting for and catching food
2. Feeding (ingestive) behavior: chewing, tearing, swallowing, licking, lapping, etc.
3. Elimination behavior: cats burying feces, marking territories, eliminating away from nest, etc.
4. Predator avoidance or Defensive behaviors
5. Shelter-seeking behaviors: climbing trees, sheltering in coral reefs, retreating into dens, etc.
6. Construction behavior: nests, burrows, bowers, egg cases, webs, etc.
7. Reproductive behavior: courtship, bonding, copulation, egg-laying, etc.
8. Caregiving (epimeletic) and care-soliciting (et-epimeletic) behaviors
9. Warning behaviors: warning calls, signals, etc.
10. Thermoregulatory behavior: basking in reptiles, shade-seeking, worker bees cooling the hive, etc.
11. Molting/emergence/shedding behavior: exoskeleton molting in arthropods, cuticle shedding in nematodes, skin shedding in snakes, etc.
12. Hatching behavior: from an egg
13. Cleaning and self-maintenance behavior: cats, preening in birds, sea otters cleaning and oiling their fur, etc,
14. Olfactory behavior: tongue flicking in snakes, antennal flicking in lobsters, dogs sniffing the ground, etc.
15. Migration behavior: round trip journeys to feeding and/or breeding areas
16. Dormancy behavior: hibernation, sleep, diapause, torpor, etc.
17. Locomotion behavior
18. Navigation/piloting/homing behavior
19. Play and exploratory behavior
20. Social behavior
21. Contagious behavior: mutual stimulation to copy another's behavior
22. Communication behavior
23. Agnostic behaviors: aggression, conflict, defense, etc.
24. Territorial behavior
25. Tool-using behaviors: in chimps, woodpecker finches, sea otters, humans, etc.
26. Circadian and rhythmic behaviors: based on daily rhythms, tidal rhythms, moon cycles, yearly seasons, etc.

Note: Obviously some of these categories can show overlap to some degree depending on the specifics of the animals and situations. 1–13 are certainly functional categories because the function of the behavior is somewhat obvious from the name of the category, but 14–26 are not so obvious as to function, though they are traditionally discussed categories. Example: What is the function of social behavior? There are many answers depending on the animals and situations involved. Animals may group for defense, for warmth, for reproduction (leks), etc.

behavior, while others are limited to only a few. Also, feeding behavior can be totally instinctive in a spider, while learning can play a large role in the feeding behaviors of other animals such as some mammals. Even mating behavior is not instinctive in the apes (and humans!), as has been discovered when apes raised in captive isolation are thrown together with members of the opposite sex and either end up not attempting mating or being totally inept in the process.

There are some functional categories that are quite rare and are engaged in by only a few species. One striking example of such a behavior is engaged in by a particular fluke (*Leucochloridium paradoxum*) which parasitizes songbirds as its definitive host, and land snails as its intermediate host. The adult flukes live in the songbirds where they mate and produce eggs. The eggs leave the bird with its droppings. If a land snail accidentally ingests some of these eggs, which tend to end up scattered all over the environment, then they become infected with the fluke larvae which hatch from the eggs. After some growth and development in the snail, the large larval flukes move into the snail's tentacles or antennae where they pulsate their banded bodies within the stretched and nearly transparent tentacles. This amazing display attracts the attention of songbirds. The plump pulsating larvae apparently resemble tasty worms or caterpillars to the extent that birds often fly down, rip off the snail's tentacle, and swallow the irresistible fluke larvae.

The function of the pulsating movements of the fluke larva is *to get itself eaten* by a bird! This is the only way the life cycle can continue. There are other known examples of this strange category of behavior, but it is certainly rare compared to say feeding behavior or locomotion, and it is not included in Table 14.2. Undoubtedly there are many other rare functional categories as yet little-noted by behaviorists in the vast realm of animal behavior.

There can be vast diversity even within one of the complexity or functional categories. A whole book can be written covering say locomotion in animals, or reproductive behavior, or learned behavior, etc. A *partial* list of locomotion methods in animals (Table 14.3) should give an indication of the great diversity present within some of these major groupings.

Other traditional categories of behavior are neither complexity groupings nor functional groupings. Some of these groupings are also listed in Table 14.2. Circadian behaviors are behaviors that occur on a rhythmic cycle of, or approximating, 24 hours. Saying that a behavior is circadian might suggest that it is instinctive, but gives no idea of what the behavior is, how complex it is, or what its function might be. Sleep (typically a

Table 14.3 Some of the Various Methods of Locomotion in Animals

Walking/running
On 2 legs: humans, birds
On 4 legs: most tetrapods, preying mantids
On 6 legs: most adult insects
On 8 legs: most spiders, scorpions
On more than 8 legs: centipedes, millipedes, isopods, velvet worms

Hopping/Jumping: Frogs, fleas, grasshoppers, kangaroos, some lemurs, anoles

Rolling "Downhill": Some tarantulas and salamanders

Swimming
By cilia: rotifers, gastrotrichs, ctenophores, many animal larval forms
By use of tail thrust (dorso/ventral flips): dolphins, whales
By use of tail thrust (lateral flips): most fish
By use of paired lateral fins: rays, cuttlefish, some bony fish (in part)
Dorso-ventral body undulations: dolphins, whales, leeches, nematodes
Lateral body undulations: some fish (in part), sea snakes, eels
By jetting of water from mantle cavity: octopus, squid, some scallops
By jetting of water from rectum: dragonfly larvae
By contraction of the "bell": most jellyfish
By contraction of webbed arms: some deep-sea octopods

Flying
With 2 modified forelimbs: birds, bats, pterosaurs
With 4 wings of nonlimb origin: dragonflies, bees, lacewings
With 2 wings of nonlimb origin: horseflies, mosquitoes

Aerial gliding/"floating"
Ballooning of baby spiders on long silk threads
Gliding of flying fish with their long pectoral fins
Gliding of flying squirrels and similar mammals using the loose skin stretched
    from their forelimbs to their hind limbs.
Gliding of some tropical snakes by flattening their entire body
Gliding of some tropical frogs by expanded toe webbing

Burrowing
By use of body peristalsis: earthworms, some marine worms
By use of a muscular "foot": some clams
By use of limbs: moles, mole rats, some insects

Terrestrial undulatory movements: most snakes, some nematodes

Whole body shortening/lengthening with sucker attachment: leeches

Whole body bending and straightening with leg attachment: inchworms

Somersaulting: Hydra

Note: This is only a partial list which could be greatly expanded and subdivided. Some locomotion methods are especially rare and unique, like the amazing locomotion of sidewinder rattlesnakes across sand (quite distinct from the locomotion of other snakes) and the surface gliding of Planaria using its ventral cilia.

circadian behavior) is a category of behavior found in some animals. Sleep is not well defined for all animals, nor does it occur in a regular pattern across animal groups. Most importantly, scientists do not firmly know what function sleep accomplishes. It may indeed have several functions that vary from one animal group to another. Again, sleep is neither a functional grouping (yet) nor a complexity level of behavior, though it does seem to clearly represent a rather definitive category of behavior.

In short, behavior is diverse. Behavior is so diverse that it is not easily grouped and categorized into any obvious descriptive framework. The only general statement one can make about behavior is that it is usually adaptive for those organisms performing it. Behavior is certainly an obvious part of the total organism in some groups, and it is often more diverse, and sometimes less diverse, than the diversity of species within the group under consideration. It therefore represents yet another interesting and valid parameter of biodiversity.

15

Systematics: Elucidating and Representing Biodiversity

Our classifications will come to be, as far as they can be so made, genealogies.

—Charles Darwin, *The Origin of Species*, 1859

In recent years it has been gradually realized that taxonomy is not merely a necessary pigeon-holing but also one of the most important activities in biology, requiring a synthesis of all other biological pursuits for its proper performance, and producing results of the highest importance in the study of evolution.

—A. J. Cain, *Animal Species and Their Evolution*, 1993

Taxonomy provides a view of biodiversity on a scale not replicated by any other discipline.

—Quentin D. Wheeler, Systematics, Overview, in *Encyclopedia of Biodiversity*, Vol. 5, 2001

Systematics is the discipline that addresses the evolutionary relationships between and among species and species groups. Taxonomy is the discipline concerned with formally naming and grouping species into hierarchical groups like kingdom, phylum, class, etc. Before Darwin's revolution, only taxonomy existed. Today taxonomy and systematics have largely fused into one inseparable discipline, and one can't be a taxonomist without taking into account what is known about evolutionary relationships. The grouping of species into "higher" taxa such as genera, families, orders, etc. is today based mainly on the relative closeness of their evolutionary relationships (their phylogeny).

Of all the subdisciplines of biology, systematics more than any other concerns itself with formally making sense out of biodiversity and representing it in a consistent manner (for the most part phylogenetically).

Though a few biodiversity parameters are not addressed by systematics, many such as species diversity (Chapter 3), higher level or deep diversity (Chapter 5), morphological diversity (Chapter 9), genetic diversity (Chapter 6), and even metabolic diversity (Chapter 10) all figure into the pursuit of modern systematists.

Systematics is a form of categorization. It seems logical that man's first steps in the direction of science were in the area of categorization, that is, grouping things according to some set of characteristics. Some animals and plants came to be considered "food" while others were not. Some were considered to be dangerous while others were not. Some types of rock were easily chipped into stone tools while others were not. Different members of early human groups came to be known by their unique personality traits such that they were divided mentally into friends/enemies, good hunters/poor hunters, or trustworthy/untrustworthy. Because of the complete reliance of primitive peoples on the living communities surrounding them, they also classify most of the surrounding "species" according to some set of criteria that allows them to use those species names for communication of important information concerning their environment. Through their early categorization of the surrounding world, humans started to make sense of their environment. A knowledge base was then in place upon which further understanding of the world could be constructed. Our current knowledge base concerning life on earth has become vast and highly interconnected through synthesis, and of course it is still growing.

Today we understand that no useful scientific information can be obtained in biology unless you know precisely what biomolecule, cell type, system, or species the work is based on. A field biologist can't simply say that his 3-year field study on birds involved "geese." He must state which of the many species of geese he studied, and unless the study was of a comparative nature, he should stick to only one species at a time to obtain data that are accurate descriptors for that species. An immunologist could not say the "small brown monkeys" used in my study were innately resistant to the anthrax bacteria. This tells us essentially nothing because there are many species of small brown monkeys, some of which may very well not be resistant to anthrax. Accurate classification is crucial to any detailed understanding of the living world because even closely related species may vary greatly in some traits.

Though Aristotle (384–322 BC) was one of the first formal classifiers of nature, biologists generally go back only to Carolus Linnaeus (1707–1778) as the person most responsible for the framework of modern biological classification. Linneaus established a hierarchical or nested system of formal grouping for both the naming and grouping of organisms. His nested grouping levels were

Empire	Everything to be classified
Kingdom	He had three: animal, vegetable, and mineral
Class	He listed far fewer under plants and animals than we have today
Order	Again fewer per Class in most cases
Genus	Again fewer and broader than modern genera
Species	Similar to today's usage, but recognized entirely by morphology
Variety	Similar to clines, subspecies, or races of today

These levels represent *ranks* with each lower rank representing a group which is united by more shared characteristics than the group(s) above. That is to say, two species in the same order would share a greater number of characteristics in common than either would with a species from another order. This is still largely true in our current classification system, though some parts of the system are constantly undergoing review and revision to perfect their accuracy and usefulness (a sign of healthy growth and refinement in science).

One of Linnaeus's greatest contributions was the idea of *binomial nomenclature*, by which each species is referred to using both its genus name *and* its species name as in *Homo sapiens, Amoeba proteus*, or one of my favorites—*Gorilla gorilla*. According to the rule of binominal nomenclature, no two species can have the same genus *and* species name. Using this rule scientists have been able to handle and accurately deal with species-level biodiversity. Later workers increased the usefulness of the Linnaean system by removing empire (which was really unnecessary) and adding some additional group/rank levels. Today we hold that there are seven *major* (and necessary) nested categories of grouping (refer back to Table 5.1).

As you can see in Table 5.1, the phylum and family levels are the newer additions. Any living organism is required to have a name at these seven levels of classification, even those which are the only members of their phylum, such as *Trichoplax adhaerens*, which is the only species known from the animal phylum Placozoa. In especially diverse groups, the required seven rank levels have not been enough to effectively group and subdivide the large array of resident species, so new levels have been created and inserted as needed within the main seven.

Most of these are listed below along with the original required seven (underlined).

Kingdom
Subkingdom
Grade

Phylum
Subphylum
Superclass

Class
Subclass
Cohort
Superorder

Order
Suborder
Infraorder
Superfamily

Family
Subfamily
Tribe

Genus
Subgenus

Species
subspecies

Though many of these rank levels are little used, they have been deemed necessary within certain diverse groups like the Arthropods. More levels may be formally added in the future such as the three domain levels mentioned in Chapter 5 (see Figure 5.2), which are already presented in most new general biology textbooks. Of course new species are discovered almost daily, and they are added to our classifications at appropriate places. Some of these new species not only count as new species in already existing genera, but are also distinct enough to be counted as new genera, families, orders, classes, and even phyla. In 1995 a new species of marine animal (*Symbion pandora*—see Figure 3.1) turned out to be so different from all other animals that it was judged to be a new phylum of animals (Cycliophora). These additions are all welcome, as they expand our view and understanding of biodiversity. We are still building our understanding biodiversity, and more understanding will likely require additions and modifications to our taxonomy, as it always has.

Some workers are now arguing for much more than continuing modifications of our current taxonomic system. They envision a complete revolution that would result in many more kingdoms, less emphasis on rank status, and many more names between kingdom and species levels—among other proposals. They argue that some groups like the plants are known to have such a large number of branching clades that the current system just isn't up to the task of containing and accurately representing plant phylogeny. Though these "radicals" are still in the minority, they do raise important points and arguments which stimulate the thinking of all biologists and systematists; again—a healthy state of affairs.

Linnaeus grouped mainly by morphological similarity and in part by habitat. He had no notions about evolutionary or lineage relationships between his groups. Darwin's revolution slowly changed all that. We have retained the hierarchical nested groups of Linnaeus, but modern systematics attempts increasingly to group species by phylogeny as well as by morphology (see Darwin's quote leading this chapter). A short-lived movement called "Phenetics" revived the idea of classifying organisms mainly on the basis of morphology, especially in botany in the 1960s–1970s, but it has since died a relatively rapid death, and you would be hard pressed to find a modern biologist who would admit to being a pheneticist.

Most taxonomy today is the result of a mix of *evolutionary taxonomy* and *cladistics*. Cladistics (or phylogenetic systematics) is a view which is the extreme opposite of phenetics. In cladistics, classification is a matter of phylogeny—the actual evolutionary relationships of species and species groups. For the cladists, morphology is only one of several sources of information, and in some cases its relative importance may be meager. The morphology of some adult parasitic crustaceans is so aberrant that not only do these creatures not look like crustaceans, they don't even resemble arthropods—the phylum which contains crustaceans (Figure 15.1). In this case, evidence from larval stages, development, biochemistry, genomics, etc. was used to place these weird crustaceans in their respective orders, families, and genera to show their evolutionary positions regardless of their atypical adult morphology.

Evolutionary taxonomy, though it sounds like it might be synonymous with phylogenetic systematics (cladistics), lies somewhere in the middle ground between phenetics and cladistics. Evolutionary taxonomy recognizes the importance of utilizing phylogeny in classification, but it also recognizes major morphological distinctions between certain "groups," even when those groups don't correspond to phyletic/cladistic groupings. An often-used example of how evolutionary taxonomy and cladistics differ is in the current class Reptilia. Reptiles seem to form an obvious grouping of animals due to several morphological characters which are shared among its members. We speak of the reptile "grade," meaning the suite of characters including scales, ectothermy, nucleated red blood cells, certain unique skeletal details, etc. Evolutionary taxonomy recognizes this vertebrate grade by grouping these animals under the class Reptilia. Cladists, on the other hand, refuse to recognize reptiles as a taxonomic group because they do not form a complete clade. From within the "reptile" ancestry sprung (separately) two other vertebrate grades, the birds and the mammals, which are also classified as Classes by the evolutionary taxonomists. According to the cladists, the reptiles, birds, and mammals form one single large clade which should be given a formal taxonomic name. The birds and mammals can retain their clade names because they are smaller separate clades, but 'the reptiles" are not a complete clade, though certain reptile groups like the turtles and snakes do represent clades, and could be given appropriate names (Figure 15.2).

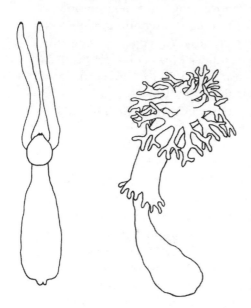

Figure 15.1 Two adult marine parasitic copepods of the phylum Arthropoda: Subphylum Crustacea—*Ommatokoita elongata* (left) and *Phrixocephalus longicollum* (right), neither of which resemble typical crustaceans or arthropods. The morphology of *Phrixocephalus* is so extreme that it is not even clearly recognizable as an animal. (*Figure by author*)

A "group" such as the reptiles which represents only part of a clade is said to be paraphyletic, and cladists do not accept paraphyletic groupings as valid taxonomic or systematic groups despite the fact that they might "look" like they belong together. The problem, according to the cladists, is that many of the higher taxonomic groupings that were established by the evolutionary taxonomists are in fact paraphyletic. In the five-kingdom system, the Monera are paraphyletic because they gave rise later on to the Protoctista. The Protoctista are paraphyletic because they gave rise separately to the fungi, plant, and animal kingdoms. In

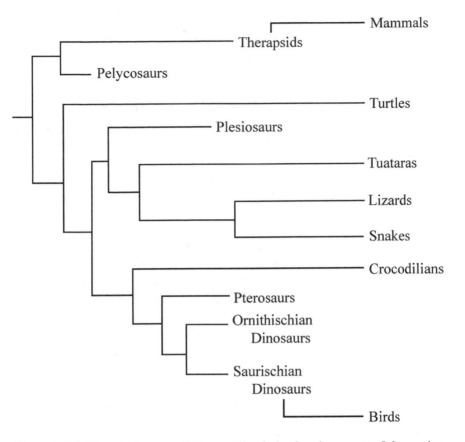

Figure 15.2 The phylogeny of the reptile clade showing most of the major groups, including the birds and mammals which have been traditionally excluded from the Class Reptilia by evolutionary taxonomists and placed into their own respective classes. (*Figure by author after various sources*)

the animal kingdom the class Osteichthyes (bony fish) is paraphyletic because bony fish gave rise to the class Amphibia. Amphibians are paraphyletic because they gave rise to the class Reptilia, which, as mentioned earlier, is itself paraphyletic. Obviously, some major aspects of our current classification system are not arranged or named in accord with phylogeny as the cladists desire.

If you still don't catch what the arguments and problems are, here is an analogy which will help with the problem of paraphyletic groupings. Recognizing a paraphyletic group would be like a couple that had four daughters, three of which grew up to be "good" daughters in the eyes of the parents. The fourth daughter, however, grew up, moved away, became a drug addict and joined a strange cult. The infuriated parents decide to disown her, remove her from their will, and no longer consider her a family member. The parents have defined their family to exclude one daughter who *was* their biological daughter. The parent's morals and beliefs which led to the disowning of one daughter are subjective, as are all value judgments. The biological reality is that the wayward daughter is still a member of her parent's family and should be objectively recognized as such. Cladists say that evolutionary taxonomy subjectively places too much emphasis on a few morphological characters such as scales to define reptiles, while disowning two other members of the clade for their "wayward" morphologies involving feathers (birds) and fur (mammals).

Another interesting example for understanding the cladist position is the true statement that a *Tyranosaurus rex* (class Reptilia) would be more closely related to a pigeon (or any bird—class Aves) than it is to a fence lizard (another reptile). Since birds arose from the same reptile branch as the *T. rex*, birds share a more recent lineage with *T. rex* than *T. rex* does with lizards, which themselves shared a common ancestor with *T. rex* much further back in time (more generations removed from *T. rex and* birds—Figure 15.2). So, the way it stands now, we have one reptile (*T. rex*) more closely related to all the animals in another class (Aves— the birds) than it is to many other members of its own class, which for the cladists is simply unacceptable.

Cladists argue that giving great weight to obvious morphological grades is a biased form of classification, as we tend to be biased toward noticing the macromorphology of organisms. Morphologies though can diverge drastically even among closely related forms, as was mentioned earlier for some parasitic crustaceans. As we have seen morphologies can also converge from unrelated lineages (see Figures 1.1 and 1.3). The cladists argue that in several cases these problems make classification by morphology a subjective, rather than an objective exercise, and of course science in any form strives to be objective. Through numerous channels of evidence, cladists believe that we can eventually determine the true phylogeny of earth's life-forms and classify them according to this one objective measure—their phylogenetic relationship.

One can only wonder which school of taxonomic thought Darwin would side with were he alive today, but if he were true to much of what he wrote in *The Origin*, he would probably favor the cladistic line of thinking. The arguments and discussions between the evolutionary taxonomy and cladistic schools of thought are ongoing, but it would be accurate to say that our current system of evolutionary taxonomy is slowly but surely incorporating more cladistics into its structure. During the last 30–40 years we have moved from a period where we thought or at least pretended that we understood phylogeny, to a period where we admitted we didn't really know much about phylogeny, to the current period where we still admit our ignorance while at the same time making great strides toward understanding phylogeny on a grand scale. Such advances often affect the taxonomic arrangement of the organisms involved. Listed below are a few examples of this progress.

- Domains: The concept of three domains was essentially unknown prior to 1996; now it is in even basic biology texts. This "tree of life," as it is sometimes called, is based mainly on comparisons of certain regions of ribosomal RNA molecules, but other evidence from other sources is now being gathered to edit and refine this revelation of life's history and relationships.
- Kingdom of the Fungi recognized: Only three to four decades ago some workers still felt that fungi should be grouped with plants, as Ernst Haeckel had done with his three-kingdom system in 1866.
- Kingdom Protoctista (Protista): Due to increasing evidence from ultrastructure, biochemistry, physiology, etc. the number of phyla in this kingdom has gone from around seven to eight in the 1960s to around 25-plus currently—indicating again that these groups were not close relatives and deserved separate phyla status. Some workers even feel that these very old branches of the tree of life deserve to be kingdoms in their own right. Many of these groups separated or diverged long before the plants, animals, and fungi did. Only time will tell how all these ideas settle out.
- Phylum Aschelminthes: This was once an accepted phylum of animals said to be united by a type of body cavity called a pseudocoelom. Now this group has been split into several separate phyla (Nematoda, Nematomorpha, Rotifera, Acanthocephala, Kinorhyncha, Loricifera) because other evidence indicates that they do not form a single clade, and the pseudocoelom is believed to be convergent in some of these phyla rather than homologous.
- Class Agnatha: This class included the hagfish and the lampreys as the two examples of living jawless fish (Figure 15.3). Now the two groups have been split into separate classes, the Myxini and the Cephalaspidomorphi respectively, because they turn out by developmental, ultrastructural,

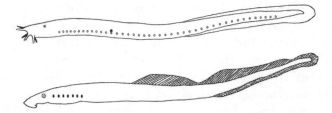

Figure 15.3 A hagfish (above) and a lamprey (below). Previously these two types of fish were grouped in the Class Agnatha based on their similar morphology and the assumption that they were closely related. New evidence, including molecular data, showed that lampreys were closer to sharks and rays than to hagfish, so the Class Agnatha was abandoned and the two groups were split and placed into separate classes: Myxini for the hagfish, and Cephalaspidomorphi for the lampreys. (*Figure by author*)

and biochemical evidence not to be that closely related (even though their morphology might place them in a common grade). One major distinction now recognized is that while lampreys are vertebrates (they do have small "vertebrae" of cartilage), hagfish lack vertebrae and probably never had them.

There are still other examples, but the point is that our system of classification continues to change with new information from all levels of investigation: biochemical comparisons, fossil record, details of development, metabolic comparisons, studies of ultrastructure such as cell anatomy, and the discovery of new species (which sometimes fill important gaps in our understanding of phylogeny). It is an exciting time for those interested in the work itself, or in the outcome of the many recent and ongoing phylogenetic studies that are revealing much about how biodiversity evolved and how the many diverse groups of organisms are related.

Though scientists may resist drastic changes in how organisms are classified, perhaps such change will be necessary to free us from our preconceived notions about biodiversity and what subjectively chosen characteristics define groups, notions that derive from our current standard system of classification. In her excellent 1998 book Symbiotic Planet, Lynn Margulis wrote: "Our classifications blind us to the wildness of natural organization by supplying conceptual boxes to fit our preconceived ideas." Some species and groups for instance, have a strange mix of fungal and protoctistan characters, and may be as distinct from either group as fungi are from "true" protoctists. The various phyla of both prokaryotic and eukaryotic unicells are largely as unrelated to each other as squirrels are from mushrooms. Perhaps moving to 30 or so kingdoms would go a long way to opening our minds to the extreme diversity of biodiversity, a view of diversity that is harder to see from a system of only five kingdoms.

16

Biodiversity and Values

In this day and age, it is clearly important to be biologically literate, but lest we forget, the study of life is also an aesthetic experience.
—Daniel Chiras, *Biology: The Web of Life*, 1993

The learned is happy nature to explore, The fool is happy that he knows no more.
—Alexander Pope, *Essay on Man*, 1732

Let me give you a definition of ethics: It is good to maintain life and further life; it is bad to damage and destroy life.
—Albert Schweitzer, The Relation of Love, in *A Treasury of Albert Schweitzer*, 1965

We love the things we love for what they are.
—Robert Frost, *Hyla Brook*, 1920

As stated earlier, and as you may already be aware, most of the books and writings on biodiversity are in large part concerned with two biodiversity issues. One is the pragmatic importance of biodiversity to us and to the health of the planet, and the other is the fact that we are in the midst of, and are largely responsible for, the earth's sixth major extinction event. In short, the message is that biodiversity is very important and, at the same time, very endangered by human populations and activities, and both of these conclusions are certainly well justified.

One question to be raised is—just how widely known and appreciated are these important facts concerning the importance and decline of biodiversity? Sadly, many do not know them well enough because neither their education nor the media have done an adequate job of filling them in on the existence and importance of these hugely crucial issues. Ralph Nader has spent most of his adult life trying to make people more responsive to the important collective issues and needs of society, and though he has certainly had an important effect, it has not been nearly as large as he would wish. Most people everywhere are simply too selfish and self-concerned to care much about the immensely bigger issues of environment, responsible government, national health care, consumer safety, etc. Others may care but fail to act because they believe they can have no effect on such huge and complex issues. Even when we are well aware of these large and important problems, our response is largely one of mere talk and complaints, and even those are not properly directed to those who could have an effect on the problem. Most of us are guilty at least to some degree, and this author is to be included among that number.

Other animals presumably have no concept of biodiversity, nor any concern for the living realm beyond their immediate (in both space and time) existence. Animals are the only organisms with nervous systems and brains which could possibly entertain any thoughts, feelings, or concerns for biodiversity, and only a very few animal species have brains large and complex enough to process and reflect on what we call "thoughts." Even among our closest and brightest relations, the primates, animal lives are lived for the most part "in the moment," with little thought of the past or future. Even nonscientists have mentioned this idea, including the famed Scottish poet Robert Burns, in these lines from his poem entitled *To a Mouse*.

> "Still, thou art blest, compared to me!
> The present only toucheth thee:
> But Ouch! I backward cast my eye
>   On prospects drear.
> And forward, though I cannot see
>   I guess and fear."

Some biologists believe there are exceptions—at least in some of the apes and possibly in the whales and dolphins. Still, it is doubtful anyone could make much of a case for other animals valuing other species (or biodiversity in general) where those species don't represent food, raw materials, immediate stimulation, etc. Probably only humans have the ability to aesthetically appreciate the beauty of a tiger or a butterfly, to feel a sense of awe at the range of diverse creatures on the planet, to have reverence for life, or to feel in an almost spiritual way that we are a connected part of the whole of life on earth through our common ancestry.

Values are not objective phenomena, they are subjective opinions, concepts, and beliefs. Values do not exist "out there" in nature. They only exist as ideas in the minds of thinking beings. Humans are a diverse species in many ways, including our values. Most humans are individually pragmatic and selfish, exactly like most other animals on the planet. Some humans do give of their effort, time, and money out of empathy for other humans. Far fewer do the same for other species or for nature in general. Feeling real concern for biodiversity is probably an acquired value that comes either from formal education or from living closely with nature, as many of the native peoples of the world did until recent times (a very few still do).

Without question, most people today are not truly aware of the existence and magnitude of the biodiversity crisis, nor are they aware of why it matters. Education is certainly part of the solution, but how is that to be achieved? Many, if not most, college graduates in the United States have never had a course designed to make them aware of the concept of biodiversity, the decline of biodiversity, or the value of biodiversity to humanity. If they take a course in philosophy, which some do, they would dealt with values, but mainly with values that revolve around the human species and the possible relation of such values to metaphysical belief.

Obviously courses could be developed which attempt to educate students about biodiversity issues, including those of value and importance. The question is whether universities would vote such courses into their general studies or core curriculum such that most students would end up taking one. Many universities today are cutting back core curriculum requirements and allowing more "student choice" in their programs. The climate is not generally favorable to adding a new course and discipline into the required, or even recommended, list of courses. It is easy and tempting to be pessimistic, but if we don't try to make people biodiversity literate, it won't happen. Only when people become more informed concerning biodiversity issues will there be a force for change, one which could slow or even stop the current decline of biodiversity. Some of that change could result when knowledgeable citizens vote for candidates who speak to and support (both in word and action) environmental conservation and preservation.

Some 150 years after Darwin's revolution reshaped biology and our view and understanding of our place in nature, millions of otherwise educated people still reject his Herculean synthesis of life's process and history (another failure of our educational system). The same rejection, or at least apathy, is apparent in relation to the biodiversity crisis among large sections of our modern educated societies. For many, the value of jobs, territory, economic development, religious belief, recreation, etc. overrides the incomprehensibly greater value of the natural world. Like biodiversity itself, the *value* of biodiversity is currently more remote and less comprehensible for many people than the simple and immediate value of a job, a strong stock market, or the pleasure of their favorite recreational activity.

Over the last few decades, many of the "developed societies" have seemed to move more and more toward the ideals of self-fulfillment, hedonistic pursuits, selfishness in terms of both possessions and behavior, and consumerism on a scale that is beyond all reason. Of course this is not a blanket statement, and individuals will certainly pull together in emergencies like the September 11th event in New York City (2001). Still, much of the media engages millions with tantalizing displays of grandiose wealth, hedonistic behavior, shopping channels, and advertisements for largely unnecessary products that take up a sizable portion of network airtime. Anyone who samples the multitude of available television channels would have to agree.

There is clearly not enough exposure of biodiversity issues in modern society. Even among those who enjoy watching specials about science, nature, animals, etc., too many want to see only the beauty, without being reminded of the problems. We all need to be better educated about the ongoing environmental crisis in a way that would influence us to take more active roles in slowing or stopping the destruction. As mentioned earlier, even students working on degrees in biology are not much exposed to the accelerated state of declining biodiversity (the raw material of biology) now in progress.

Some of the values of individuals have an objective basis, as in one who values the music of Beethoven, the films of Stanley Kubrick, or the writings of Loren Eiseley. Such personal values have come through experience and through active choice of valued works from among other experienced works, and yes—larger groups can share similar values of opinion. But the "higher" values we hold are more in the realm of ideas than of objectively experienced phenomena. Very few values are shared universally except the values of food, water, shelter, sex, children, and others of obvious biological importance. Not all people especially value (or understand) individual freedoms as those in the United States do. Not all believe in and value the concepts of human equality, freedom of speech, religion and religious freedom, the golden rule, charity, "family values," etc.

Of all the values we hold collectively or individually, one that many do hold is that people *should* make the most of their talents and abilities. This is sometimes stated in the reverse—people *should not* waste their talents and abilities. Educators and many others believe that people should never stop learning, questioning, and exploring their world, because these are precisely the abilities in which humans excel, if these talents are nurtured rather than stifled. Another talent we humans possess is the ability to find value in phenomena which are far greater than ourselves. We can value humanity as a whole. We can value other species both singly and collectively. We can value that vague yet familiar phenomenon we call Nature—the living world.

It is we alone among the billions of species that have ever existed who can both know *and feel* that we are connected biologically and historically to the immense whole of earthly life. It is we alone who can feel the "grandeur in this view of life." This phrase from the Darwin quote back in the preface has spoken to many in the biological sciences and to those in the general populace who have some understanding of evolution and the connected unity of life. Though environmental decline was not as severe or obvious in Darwin's day, it is probable that were he alive today he would be at the forefront along with E. O. Wilson and others who are attempting to make us all more aware of biodiversity, more knowledgeable about its importance and value, and more alarmed at its decline as the direct result of the varied global activities and growth of the human population.

If we possess this wonderful and enriching ability to value the whole of earthly life, should we not be developing it to the fullest? Should not our educational institutions make this development a core element of any liberal education? It cannot be argued that this value is mere sentimentality, for our very existence depends on intact functioning ecosystems. The oxygen we breathe and the food we eat is produced by this living world. Life is collectively the most complex set of entities known to exist in the universe. Life has a history on this planet that is incomprehensibly older and more complex than our recent and brief residence. Life includes "us." These should all be obvious reasons for valuing biodiversity that any human could understand. Add to those the awe and pleasure that can be derived from increased understanding and comprehension of biodiversity and the argument for promoting increased awareness becomes even stronger.

Perhaps there are no universal values. The question has been argued by philosophers, theologians, and others for hundreds of years with no consensus in sight. The scientific stance has always been that knowledge and increased understanding/comprehension are unquestionably "good." A side benefit of this stance is that the more we understand, the more there is to appreciate and value. Though knowledge and wisdom are not synonymous, it is doubtful whether we can ever be truly wise without knowledge. Knowledge and comprehension of life/biodiversity simply must be part of the equation for the future of humanity on this planet, and valuing biodiversity should be a logical consequence of this comprehension. If we come to better comprehend and value biodiversity, surely we could learn to interact with it more wisely and with less destructive shortsightedness. There is always hope.

17

Comprehending Biodiversity

Nothing is great but the inexhaustible wealth of nature. She shows us only surfaces, but she is a million fathoms deep.
> —Ralph Waldo Emerson, Resources, in *The Complete Works of Ralph Waldo Emmerson*, Vol. 8, 1903–1904

Man finds one of his ultimate fulfillments in comprehension. Fuller comprehension is one of the basic duties (and privileges) of the individual.
> —Julian Huxley, *New Bottles for New Wine*, 1957

My trusty old Webster's Dictionary, in defining the word "comprehend," gives: "to take into the mind; to grasp by understanding; conceivable by the mind; intelligible." Back in Chapter 3, the example of a fast-paced, but nearly year-long slide show was mentioned as an aid to illustrate the great number of species known to exist on our planet. Even if the number is as low as 3,000,000, can anyone personally understand or conceive of that number of species actually existing out there in nature? It is easy to conceive and visualize 10, or 100, or even 1,000 species (try listing a thousand species sometime—from memory!), but 1,000,000, or 3,000,000, or 12,000,000 are beyond the abilities of most people in terms of true comprehension. It is doubtful whether the human mind is up to the task of fully comprehending such numbers, though some might argue otherwise.

Along with this large number of species, consider their diversity of subspecies, genes, cells, cultures, life-cycle stages, metabolic pathways, behaviors, sensory modes, ecological interactions, and the other varying parameters we've covered in this book. Personally, this author is not up to the task of vividly comprehending it all in terms of a unified mental image of the total extent of biodiversity on this planet. Perhaps I am just more limited in my comprehension abilities than others, but I doubt it. Even if some exceptional person could truly visualize this diversity, they would have to admit that it is vast on a scale that no other known phenomena can match.

Scientists enjoy comprehending as much of a topic as possible, but they (like most other people) do appreciate the mystery still left in our world. For biologists and many other scientists, biodiversity in all its guises is a source of much that is still mysterious. Mysteries are those things that are as yet unexplained, but they are also those things that are beyond our comprehension, even when much of the complex and diverse whole is "in its parts" known and explained. Some nonscientists have expressed the sentiment that the more the world is known and explained by scientists, the less mystery there is left in the world. This may sound obvious, but there is a rebuttal to this stance. George Wheeler (a renowned physicist) wrote: "We live on an island of knowledge surrounded by a sea of ignorance. As our island grows, so does the shore of our ignorance." You may recall that this quote appeared much earlier in the book, but it is such a gem, and it so clearly states the obvious fact that there is no foreseeable end to the unknown.

After some 30-plus years of studying biology, I still stand in awe not only of what I have learned about biodiversity, but also of the many things that are yet unknown and even "mysterious" (to me personally or to the scientific community). Let me share with you what is one of the greatest mysteries I have encountered. In the oceans of the world, there are countless tiny animals known as larvaceans. They are actually among our closest relatives in the living world, though you would never guess it by looking at them. These tiny animals only millimeters long look something like a tadpole, with a rounded main body, and a wiggling flattened tail. Most of these simple-looking animals, with no obvious means to do so, can construct from their own secreted mucus a large structure around themselves called a "house" (Figure 17.1). This house is not a simple blob of mucus, rather it is an elaborately complex and symmetrical construction which includes a cavity for the animal and connecting incoming and outgoing water channels through which water is moved by the undulating actions of the tail. Larvaceans are filter feeders that feed on bacteria, diatoms, algae, etc., which are pulled in through two large surface screens (that keep larger particles out of the house). There are two internal lateral food traps of complex construction (still not totally understood) which actually collect the food and channel it toward the animal's mouth. Finally the filtered water is channeled toward an excurrent pore. As it is pumped out by the

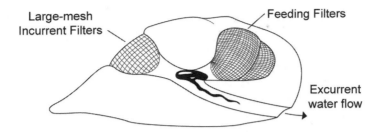

Figure 17.1 A tadpole-like larvacean (in black) within its mucus "house." The tail drives a water current which pulls water in through the forward incurrent filters, then around the animal to the fine-meshed feeding filters near the rear, and finally out through a narrow channel within which the tail is beating. (*Figure by Jeff Dixon*)

undulating tail, the water jet slowly propels the larvacean and its house through the water.

The mystery here is how could a simple tadpole-like animal construct from its own mucus such an elaborate structure? Observations of the process reveal that it takes only a minute or two from beginning to completion. The animal is seen turning back on its tail and "wiggling" during the construction, then suddenly it stops and starts undulating its tail—the house is finished. I don't know if anyone is studying this amazing feat, but it impresses and mystifies me as much as anything I am aware of in nature. The abilities of this simple creature seem to border on the unbelievable. I should add that larvaceans break out of their old houses every few hours (because they have become clogged with debris) and construct a new one with ease.

Even with its inherent mystery, biodiversity needs to be better comprehended by everyone if we as a species are to find value in it. The more we can comprehend of biodiversity, the more we may appreciate, stand in awe of, and value it for what it is, rather than for what it can do for us. Many humans have expressed the idea and ideal that we should value all people, not for who they are, or what they have done or may do, or for what they may do for us individually, but simply because they are one of "us"—a part of humanity. Of course some may argue as to whether every person deserves the same degree of value and respect, but this idea of innate human value is at least an important point of reference in most civilized societies. There are both rational and metaphysical arguments that support such an ideal.

From a similar stance, why not include all of life's diversity in this idealistic view? All species have proven themselves through 4 billion years of evolutionary history to be successful life forms, and we humans are but one among those millions. Might it not also be rational to respect and value our fellow creatures simply because they exist? Of course we should be able to alter nature where necessary to alleviate hunger, disease, and other serious obstacles to a humane life for people, but we should (must?) understand what we are doing so that we may limit the damage we do to the biosphere of which we are a part. Many wise and knowledgeable people agree that an essential first step is to recognize the simple fact that there are already far too many humans on the earth. Biodiversity simply cannot persist in the face of continued increases in human numbers, especially if we want all humans to have shelter, temperature control, adequate food, and other essentials of life that allow for the possibility of happiness.

Though most biologists do not hold with the earlier and more radical versions of the Gaia Hypothesis (that the whole biosphere is in a sense an integrated superorganism), if one briefly entertained this notion, the human species could only be viewed as a growing cancer within the living biosphere. Cancers grow selfishly at the expense of the body they dwell in. They destroy functioning cells and tissue and replace them with more cancer tissue that in no way contributes to the well-being and homeostasis of the body. This is in essence what our species as a whole is doing to the total biodiversity of this planet. We are destroying habitat and biodiversity that was relatively balanced and self-sustaining due to the cycling and cleansing processes that result from the myriad interactions of the resident species. Being compared to a cancer is not a very flattering sentiment concerning our species, but it is clearly an arguable viewpoint and analogy.

It is a hard thing to understand the destructiveness of our species, and at the same time have some comprehension of the beauty, complexity, and diversity of life that is being destroyed. It is perhaps like being a lover of literature and great writing and watching as the Nazis burned thousands of great books on their bonfires. It is very easy to be pessimistic. How can this tide of destructiveness be halted? It should be obvious that nothing—absolutely nothing—will change until more of humanity comes closer to an appreciative comprehension of our earth's biodiversity. I humbly hope that this meager work will contribute something toward that goal.

Glossary

Alleles. Variations of a single gene, which while they are said to code for the same trait, such as flower color, they may code differently for that trait—as in red flowers versus white flowers.

Archaea. One of the three domain groupings of life. It is made up of organisms with prokaryotic cells, but which differ in numerous ways from the bacteria.

Autotrophic. Refers to organisms that are able to manufacture their own supply of energy-supplying organic molecules, especially carbohydrates. Plants do this through photosynthesis. Some bacteria do this through chemosynthesis.

Character displacement. When a trait or "character" common to two similar species diverges under the influence of selection due to competition between the two species in an area of range overlap.

Clade. A group of related species of any size that includes all the species descended from a common ancestral species, including that ancestral species.

Cladogram. A visual representation of the phylogenetic relationships between the member species (or higher taxa) of a clade.

Competitive exclusion. The concept that two or more species cannot coexist in the same area when the two species share one or more limiting resources in common. One of the species will eventually outcompete the other(s) for those resources and so exclude them from that area.

Dioecious. Having separate sexes such that any one individual can be distinguished as either a male or a female.

Diploid.   This term refers to a cell having homologous pairs of chromosomes, where the same genes are located on each partner in a homologous pair. Such a cell therefore has two copies of each gene, one located on each of the two chromosomes in a pair. In most plants and animals, one member of each chromosome pair was inherited from each of the two parents. "2N" is used as a symbol for the diploid condition.

Ecological niche.   The complete set of environmental conditions within which a species can and does exist, including some components of its lifestyle such as how it nourishes itself (predator, parasite, etc.). Parameters of a niche include all necessary abiotic and climatic factors as well as the biotic factors necessary for that species to thrive in nature.

Endocytosis.   A process whereby a cell can form an inward-orienting pocket in its outer membrane, which will eventually bud into the cytoplasm as a small vacuole, carrying some materials (now in the vacuole) into the interior of the cell. Amoeba and sponges feed in this manner.

Eukaryota (Eucarya).   The domain of life that includes all organisms whose cells are eukaryotic.

Eukaryotic.   This term refers to cells what have a distinct nucleus, and typically other membranous structures such as endoplasmic reticulum, golgi bodies, etc. Eukaryotic cells tend to be larger than prokaryotic cells.

Eusocial behavior.   The extreme stage of social behavior found in some of the social insects wherein overlapping generations of related individuals cooperate in rearing the young, and where only some individuals reproduce while the others (workers, etc.) have other supporting roles to play within the group.

Evolution.   It is the change over several generations in the genetic makeup of populations, or genomic change in populations over time.

Exaptation.   When a new trait or character arises through the modification of a preexisting character, such as the wings of birds which evolved from the preexisting forelimbs of tetrapods.

Gaia hypothesis.   The hypothesis that the whole biosphere acts in a mutualistic homeostatic manner for its own preservation and regulation.

Gamete.   The general term for sex cells like eggs and sperm, though in some organisms the gametes all look alike and cannot be distinguished as eggs or sperm.

Grade. A term that refers to a group of organisms, which seem to be similar in their body plan and morphology. Because of their great similarity in overall morphology, the millipedes can be said to all exhibit the same morphological grade. The same could be said about birds. The term usually implies that each grade is relatively distinct or disjunct from other possible grades. Species sharing a common grade need not be closely related, as illustrated by the desert plants in Figure 1.1.

Haploid. A term referring to a cell which typically carries only one copy of each gene, or one that does not contain pairs of homologous chromosomes, as diploid cells do. The common examples of haploid cells in animals are the egg and sperm cells, though all the cells in a male wasp or ant are also haploid. Some of the algae and simpler plants are largely haploid. "N" is used as a symbol for the haploid condition.

Heterotrophic. This term refers to an organism that cannot manufacture its own complex energy-rich molecules from less complex and energy-poor molecules. For example the fungi and the animals are heterotrophic.

Holozoic. This term refers to heterotrophic organisms that ingest their food and then digest it inside their cells or bodies.

Homologous structures. Structures, or even molecules, which are shared by two or more species that have been inherited from a common ancestor. The cilia in human oviducts are homologous to those found in a Paramecium. The hair on a cat is homologous to the hair on a monkey since both were descended from a common mammal ancestor having hair. The wing of a bat however is not homologous to the wing of a bird because their common ancestor was not a winged species. Bats and birds evolved wings independently.

Monoecious. A term applied to an organism that produces both eggs and sperm, though monoecious organisms typically don't self-fertilize, rather they cross-fertilize with another individual. In short, the sexes are not separate as in dioecious organisms.

Mutation. Any alternation of the genetic material. This includes changes in individual genes, many of which will affect a trait, and also changes in whole chromosomes whereby genes may be duplicated, deleted, or moved to other locations in the genome.

Natural selection. The nonrandom differential survival and reproduction of genetically variant individuals. In short, some genetically unique individuals tend to survive and reproduce at higher rates than others due the selective pressures of the environment.

Parthenogenesis.   The development of an unfertilized egg, whether haploid or diploid. Some animals have adopted this method of reproduction in part or completely. Males can be rare to nonexistent in such species, depending on the exact type of parthenogenesis practiced.

Phylogeny.   The evolutionary history of a group of related species, most often illustrated through cladograms.

Polychaete worm.   Any of the worms in the large class Polychaeta in the phylum Annelida. Essentially all polychaete worms are marine.

Polyploidy.   A condition usually resulting from faulty meiotic cell divisions that give rise to diploid gametes. If viable diploid gametes fuse with either a haploid, or another diploid gamete, the resulting organism will have more than the two typical copies of each chromosome. 3N organisms are most common, though often sterile, while the rarer 4N organisms can be fully fertile with other 4N individuals. Even higher numbers of homologous chromosomes are possible.

Prokaryotic.   This term refers to the simplest type of cell known to occur. Such cells lack a nucleus and most of the other organelles typically associated with eukaryotic cells. Prokaryotic cells are typically very small. The bacteria and archae are composed of prokaryotic cells.

Protista (Protoctista).   Traditionally the Kingdom of eukaryotic organisms exclusive of the fungi, plants, and animals. It includes most algae, amoeba, ciliates, and a variety of other forms.

Sexual selection.   A subcategory of natural selection involving either male–male competition for access to females and/or female choice of mates from among several potential males. In some rare cases it is the females that compete for access to males, and the males who do the choosing of mates.

Sibling species.   Two or more species which, while being separate species in that they do not interbreed, nevertheless are indistinguishable (or nearly so) on the basis of morphology.

Stromatolites.   Limestone deposits produced by communities of marine cyanobacteria and other prokaryotes. Stromatolite fossils go back over 3 billion years, and there are several places in the marine environment where living stromatolites are still being produced. They range in size from an inch or two to massive columns several feet in height and width.

Symbiosis.   An intimate long-term or reoccurring interaction or association of individuals of two different species. Traditionally symbiosis consists of three main divisions: mutualism (where both species benefit), commensalism (where one species benefits and the other is unaffected by the relationship), and parasitism (where the parasite benefits and the host species is harmed). Recently some examples of parasitism have been found to occur within the same species (especially brood parasitism in some birds), yet the term parasitism is still used for such interactions.

Transposon.   A segment of DNA capable of moving itself to a new chromosome location (transposition). There are several classes of transposons, some of which can also copy themselves such that two or more copies can be created and transposed among the chromosomes. Transposons obviously can result in genetic mutations.

Zygote.   An egg cell that has been fertilized. The cell resulting from the union of two gametes.

Recommended Books

Cracraft, Joel & Francesca T. Grifo, eds. (1999). *The Living Planet in Crisis*. Columbia University Press.

Eldredge, Niles. (1998). *Life in the Balance: Humanity and the Biodiversity Crisis*. Princeton University Press.

Gaston, Kevin J. & John I. Spicer. (2004). *Biodiversity: An Introduction*, 2nd ed. Blackwell.

Groombridge, Brian & Martin D. Jenkins. (2002). *World Atlas of Biodiversity*. University of California Press.

Jeffries, Mike. (2006). *Biodiversity and Conservation*. Routledge.

Levin, Simon A., editor-in-chief. (2001). *Encyclopedia of Biodiversity*. Academic Press.

Margulis, Lynn & Schwartz, Karlene V. (1998). *The Five Kingdoms*, 3rd ed. W.H. Freeman.

Nielsen, Claus. (2001). *Animal Evolution*, 2nd ed. Oxford University Press.

Pearson, Lorentz C. (1995). *The Diversity and Evolution of Plants*. CRC Press.

Reaka-Kudla, Marjorie L., Don E. Wilson, & Edward O. Wilson, eds. (1997). *Biodiversity II: Understanding and Protecting Our Biological Resources*. Joseph Henry Press.

Smith, John Maynard & Szathmary, E. (1995). *The Major Transitions in Evolution*. W.H. Freeman.

Spray, Sharon L. & Karen L. McGlothlin, eds. (2003). *Loss of Biodiversity*. Rowman & Littlefield.

Takacs, David. (1996). *The Idea of Biodiversity*. Johns Hopkins University Press.

Tudge, Colin. (2000). *The Variety of Life*. Oxford University Press.

Wilson, Edward O. (1992). *The Diversity of Life*. W.W. Norton.

Wilson, Edward O., ed. (1988). *Biodiversity*. National Academy Press.

Index

About the Author

David Zeigler is Associate Professor of Biology at the University of North Carolina at Pembroke. He teaches a variety of courses on invertebrate zoology, zoology, entomology, parasitology, animal behavior, evolution, marine biology, neurobiology, and general biology.